Table of Contents

Preface

Introduction

- Embedded Software in Egypt
- It all started with the computer

PIC 16F84A

- Microcontroller Tutorial
- Getting Started With Microcontroller Programming
- This is your first program
- Run your first program on simulator
- Explaining Flasher Program in Detail
- Let 's speed up the rate
- Building your first circuit application
- Input to the PIC 16f84
- Introduction to Interrupt
- Lookup table Sine Wave Generation
- PWM Mood Light using PIC 16F84
- DC MOTOR PWM CONTROL
- LED Chaser : Larson Scanner
- PIC 16F84 Sound Generation
- Moway robot
- PIC 16F84 Tetris Video Game
- 16F84 POV AirText

- I love PIC 16F84 :LCD to PIC 16F84 Interface

- 7 Segment POV

- Hacking Infrared with PIC 16F84A

- 16F84A VGA Output

- 16F84 SERIAL COMMUNICATIONS

- 16F84A Spindle Motor Control

- 16F84A Frequency Counter

-

PIC 16F917

- New Microcontroller Chip

- 16F917 CCP Block PWM

- Yaw Rate Gyroscope interface to PIC16F917

- Analog Devices ADXL206 Accelerometer interface to PIC16F917

PIC 18F4550

- PIC 18F4550 Programmer - The best is getting better

- Pinguino Egypt - Do-it-Yourself PIC Arduino Clone

Renesas and PSoC

- PSoC Rocks!!

- Microcontroller and Success : A true Story

- Gyro Horizon. Renesas RX62N Kit

Arduino

DIY Projects

Preface

It was the year 2008 when I first got my honorable mention in Renesas DevCon. I then decided that I must start my own tutorial for beginners. I realized that I have something to tell people who want to learn about new tricks in that technology and I must do this through my blog Embedded Egypt.

When I started my blog I wanted to accomplish two clear goals:

To teach others and
To learning new stuff myself.

The two goals are mutually correlated. I learn new stuff and teach them to others so I can learn new things continuously.

Behind every post there is a success story. After a while, I thought that I can publish a book as guide for beginners.

Now, I can say that my dream came true and that is my book is at your hands.

All I want to say to people is: "Follow your Passion" and "Follow your Dream"

Do not give up your dream even if you had to work in other area or profession for living.

Keep working and searching for ways that you can practice your passion until you get chance to make significant benefit and then real profit from it.

If you kept learning, working on your dream and helping people with this dream you will get rewarded at the end.

Thank you for purchasing this book.

Ahmed Ebeid

Embedded Software in Egypt

Embedded software is one of the fast growing markets in Egypt. There are many multinational companies have opened software development offices in Egypt. There are also many companies established by Egyptian engineers. Also, there are many students and engineers who are working independently in software development for embedded systems.

As an Egyptian engineer, I invite more companies to invest in this growing industry in Egypt.

Here is a List of embedded systems companies in Egypt:

1Sheeld

http://www.1sheeld.com/

AlManar for Electronic Systems

http://www.almanar.com.eg/index.html

ATMEL Egypt

http://www.atmel.com/products/digital-broadcast/

Avelabs

http://www.avelabs.com

Axxceleraegypt

http://www.axxceleraegypt.com/contact-us/

BadrIT (Mobile Applications)

http://www.badrit.com/

Edge Technology – Egypt

http://www.edge-techno.com

Exneer

http://www.exneer.com/

GST Egypt

http://www.gstegypt.com/contactus

IDACO Egypt

http://www.idaco-egypt.com/index.php/contacts

Intelligent Services Solutions (ISS)

http://www.issholding.com

Mazid Labs - Egypt

www.mazidlabs.com

MEMS-Vision

https://www.mems-vision.com/orders-requests/contact-us/

Mentor Graphics Egypt

http://www.mentor.com/

MTSE

http://www.mtse.com.eg

Qoudra

http://qoudra.com/

QuinDev (Mobile Applications)

http://www.quindev.com

SilMinds

http://www.silminds.com

Si-Ware Systems

http://www.si-ware.com

Smartec-Group (Training and courses)

http://smartec-group.com

Tarabay Egypt

http://www.tarabayegypt.com

Valeo

http://www.valeo.com/en/home/the-group/global-presence.html

http://www.valeoservice.com/html/egypt/en/

List of Embedded Systems Components Stores in Egypt:

Some Places to buy Microcontrollers and Electronics in Egypt

Hamada Electronics

Naby Danial St. - Mahatat Misr - Alexandria

Robota Egypt

http://www.robota-eg.com/

Al Amir Electronics

http://www.ekt2.com/

Future Electronics Egypt (Arduino Egypt)

http://www.fut-electronics.com/

Grand Solutions Components

http://gs-components.com/

Megatronics

http://www.megatronics-eg.com/

Ram Electronics

http://www.ram.com.eg

http://ram-e-shop.com

Lampa Tronics.

http://lampatronics.com/contact-us/

Microcontroller & Embedded Systems Courses:

BrightSkies Technologies

http://www.brightskiesinc.com/contacts.php

Computek Training Center

http://www.computekeg.com/index.php/contact-us

Pi-Technologies

http://www.PiTechnologies.net

SGEC Group

http://www.sgecgroup.com/

Software Engineering Competence Center (SECC)

http://www.secc.org.eg/Contact%20Us.asp

TIEC

http://tiec.gov.eg/en-us/Centers/EIC/Pages/default1.aspx?cid=2

It all started with the computer

I started with the old computer AQUARIUS at the 80's. It had *Microsoft* Basic on it. I learned programming in Basic. I then got my Pentium 1 at 90's.

I learned *Microsoft* DOS and Windows. Then I learned C language.

I got AMD PC at year 2000 and I learned more about PC hardware.

I also knew about Microcontrollers (Microchip PIC, Atmel, and Renesas).

I programmed Microcontrollers in Assembly and C.

At 2006 I got Toshiba Satellite Notebook.

This year, (2008) I joined the Renesas HTS contest I submitted my project "Multichannel Oscilloscope" and guess what, I had the forth honored mention.

http://www.renesasrulz.com/thread/2225;jsessionid=7AFAE50480BD5D
F2D4D3F616A03277EA?decorator=print&displayFullThread=true

Microcontroller Tutorial:

I decided to start a Microcontroller Tutorial for beginners. If you like to learn about Microcontroller programming and embedded systems, this blog may be helpful.

Prerequisites:

All you need to know before start programming your microcontroller is basic programming knowledge, basic electronics and Boolean algebra (AND, OR & NOT gates ...).

And the first thing to know is:

What is the Microcontroller?

It is a smart IC that can be programmed to do some task. Unlike a typical IC which does a certain function that cannot be changed, the Microcontroller function is defined by its software code written on it. you can change the code when you want and thus change its function. *Smart* means it can decide and take actions according to its code.

So what's the difference between Microcontroller (uC) and Microprocessor (uP)??

uP needs some other devices to be able to work (BIOS, RAM, I/O Ports , ...etc) all these devices are other IC 's.

Although the uP is more general purpose than the uC, but the fact that the uC contains all the required devices in one package (EEPROM, RAM, IO Ports, ADC, UART, etc ...) overcomes the limited abilities and small instruction set of the uC.

Getting Started With Microcontroller Programming

Resources needed for a Quick jump-start:

For the simplicity of the learning process for the beginner, I'll start with the most famous Microchip PIC16F84 (which was the first uC I learned).

You need to prepare this stuff before start programming:

1- The target IC (the uC PIC 16F84A) and its Datasheet.

PIC16F84A
Data Sheet
18-pin Enhanced FLASH/EEPROM
8-bit Microcontroller

2- Programming circuit (JDM programmer) - you need to build it your self - or buy a Microchip PIC programmer.

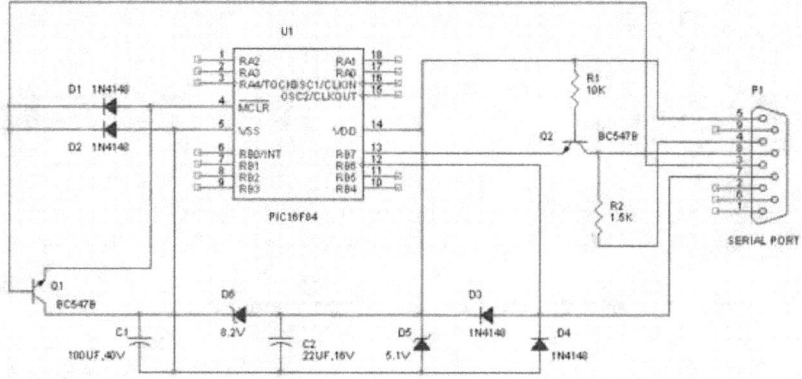

3 - Loader program (ICProg) - you'll need it if you use JDM programmer - , but if you buy a ready-made programmer, you will have the software with it.

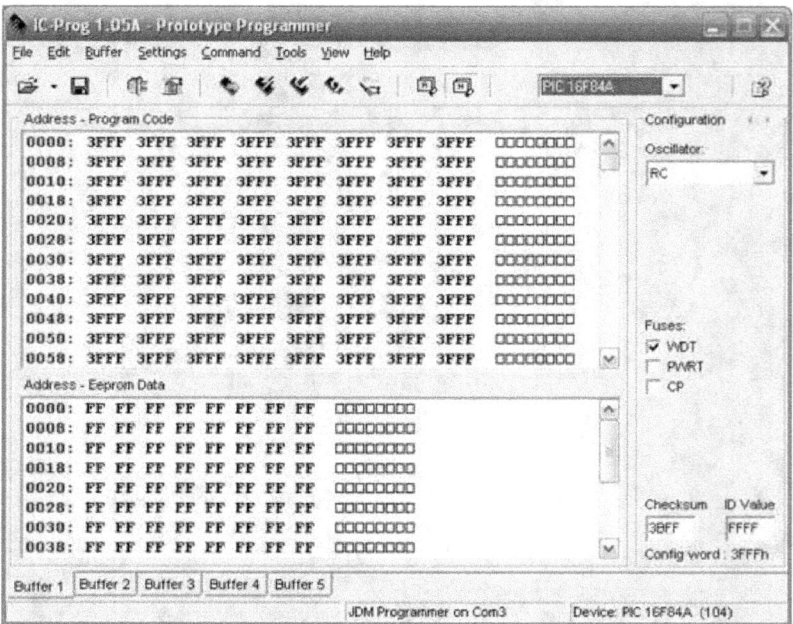

4 - MPLab program (to assemble) the programs you write in Assembly.

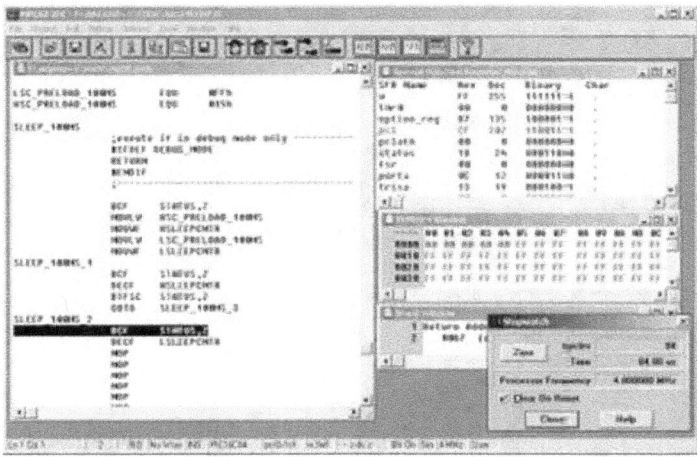

5 - HiTec PICC (to compile programs in C).

6 - Proteus 7.0 Simulator

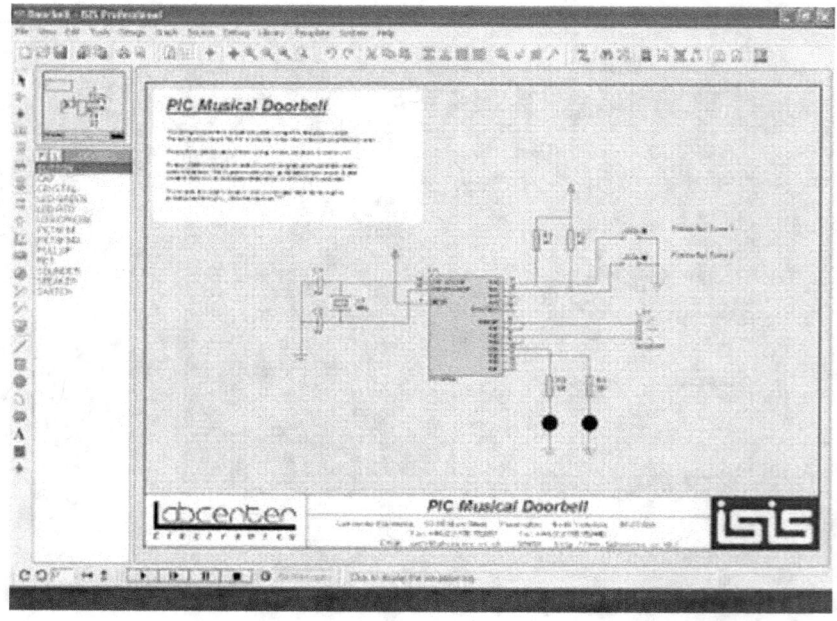

You can get all these stuff for FREE from the Internet.

This is your first program

HELLO WORLD

As in PC programming world, when you learn a new language, you start with the famous "Hello World ' program.

This program in Microcontroller is flashing a LED.

Here is the code:

```
;**********************************************
; Flasher.asm
list p=16f84                    ; This is how you can comment
include "p16f84.inc"
org 0x00
goto start
org 0x20
start
bcf INTCON,7
movlw 0x00
bsf STATUS,5
movwf TRISB
bcf STATUS,5
again
movlw 0x80
```

```
movwf PORTB

call delay

movlw 0x00

movwf PORTB

call delay

goto again

delay

movlw 0x01

movwf 0x0e

loop3

movlw 0xfa

movwf 0x0d

loop2

movlw 0xfa

movwf 0x0c

loop1

decfsz 0x0c,1

goto loop1

decfsz 0x0d,1

goto loop2

decfsz 0x0e,1

goto loop3

return
```

end

;***

1 - Copy the above text and paste it in an empty text file and save it as Flasher.asm

2 - Install the MPLab program and search for the file MPASMWIN.EXE in its path.

3- Run the program. It looks like this:

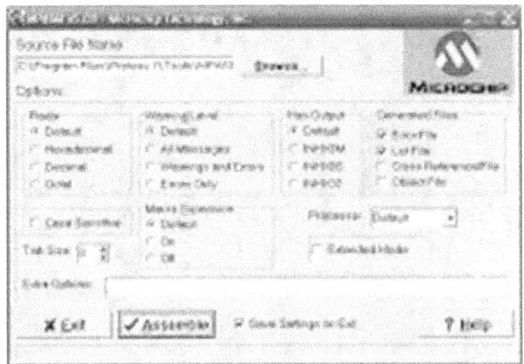

Uncheck the case-sensitive option.

4 - Browse for the *Flasher.asm* file and press Assemble.

If every thing goes write, you will have the window:

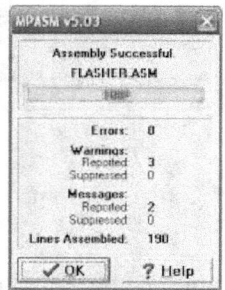

Congratulations, the assemble process is *complete*!!

The Flasher.hex file will be generated in the same directory of Flasher.asm

Run your first program on simulator

You can start to test the flasher program on the simulator

The simulator is a program that simulates Microcontroller code execution and other electronic component behavior.

One good simulator I've tried and I recommend is Labcenter Proteus 7.

You can install it and try the hardware components and software program and debug them for run-time errors and functioning errors. All this before building the real-world application on the board. This approach is very useful and can reduce developing time and frustration and eliminate situations such as you connect the circuit and get no action at all. You then wonder what's wrong. And you can get lost debugging hardware and software and don't know where to start from.

This introduction is to show the advantages of using the simulator before building your circuit board.

Let's get started..........

1- Install the simulator (Labcenter Proteus 7).

2- Run the program ISIS.

3- Draw the components of the flasher application as in this screen-shot.

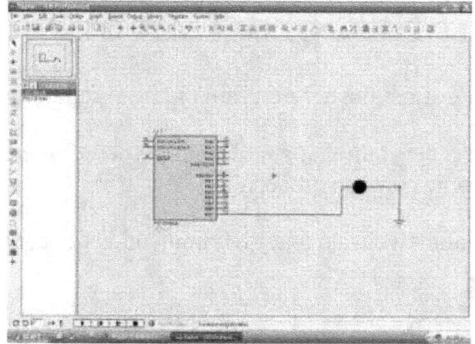

4- After you copy the code from the previous post into a file named *Flasher.ASM*, add the source file as follows;

- From the source menu, select Define Code Generation Tools...

- Choose the MPASMWIN from the scroll menu as the screen-shot,

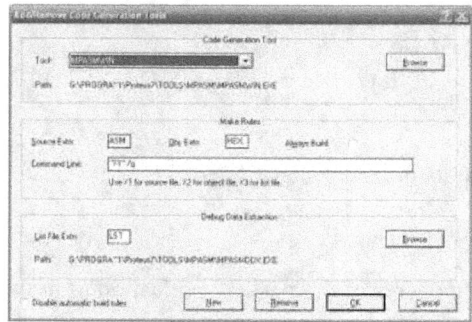

- Add the source form the source menu → Add/Remove Source files...

and choose the code generation tools MPASMWIN and the source file *Flasher.ASM*

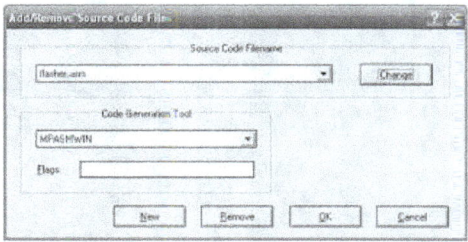

Now you have successfully configured the source file and code generation tools for code compilation.

5- You can now compile the program by choosing the Source menu → Build All.

If every thing goes right, you should get the screen

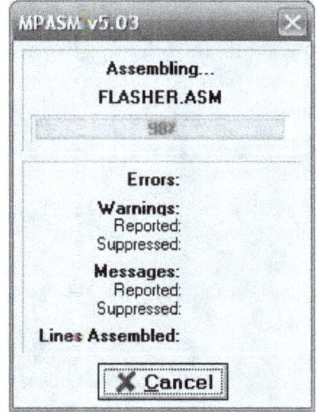

6- You will notice that a *Flaher.HEX* file has been generated in the working folder.

7- Now double-click on the PIC16F84A component in your design , you get the window,

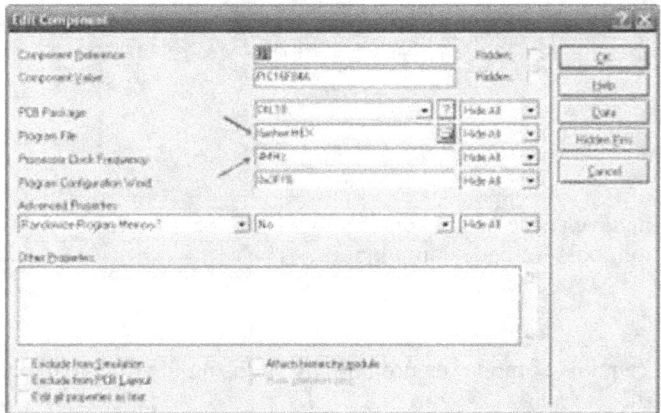

Choose the *Flasher.HEX* and set the Processor Clock Frequency to 4MHz.

8- Now press the play button to start debugging and running the program.

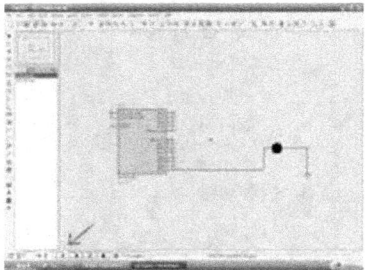

Now, you see the led flashing. Congratulations!!! The program is running.

We can start to build our first real-world circuit....

Explaining Flasher Program in Detail

You may have felt that the Flasher program is difficult. But this is not true.

Now, we'll explain it step-by-step. Just remember, when I started learning Microcontroller programming, I started by understanding this program in detail and put on it my own comments to make me remember what each piece of code did.

```
list p=16f84

include "p16f84.inc"

; This part is necessary for the compiler to know the type of the PIC you

org 0x00

goto start

; The word ORG tells the compiler to put the following code (goto

; start label) in the address 0x00 which is the reset vector of the PIC

org 0x20

start

; Again. Puts the label start at the address 0x20

bcf INTCON,7

; Clear the R7 bit in the register INTCON which disables the interrupts

movlw 0x00

; Put 0x00 in the W register
```

bsf STATUS,5

; Set the Bit 5 in the *status* register which selects Bank 1 in the RAM

movwf TRISB

; Copy the W register content into *TRISB* register [makes PORTB output]

bcf STATUS,5

; Clear the Bit 5 in the *status* register which selects Bank 0 in the RAM

again

; This is a label for the repeating part of the program

movlw 0x80

movwf PORTB

; Copies 0x80 [binary 1000 0000] in the W register

; Copy the W register content into *PORTB* register

call delay

; This command calls the *Delay* routine.

movlw 0x00

movwf PORTB

; Copies 0x00 [binary 0000 0000] in the W register

; Copy the W register content into *PORTB* register

goto again

; This label is important to keep the program running forever

delay

; This is the label for the delay routine

```
; The delay consists of 3 nested loops

movlw 0x01

movwf 0x0e

; Put 0x01 in W register

; And copy it to the memory address 0x0e in RAM

loop3

; Label

movlw 0xfa

movwf 0x0d

; Put 0xfa in W register

; And copy it to the memory address 0x0d in RAM

loop2

; Label

movlw 0xfa

movwf 0x0c

; Put 0xfa in W register

; And copy it to the memory address 0x0c in RAM

loop1

; Label

decfsz 0x0c,1

; Decrease contents of memory address 0x0c by 1 ,

; then skip the next command if the result is zero

goto loop1
```

; goto the outer loop

decfsz 0x0d,1

; Decrease contents of memory address 0x0d by 1 ,

; then skip the next command if the result is zero

goto loop2

; goto the outer loop

decfsz 0x0e,1

; Decrease contents of memory address 0x0e by 1 ,

; then skip the next command if the result is zero

goto loop3

; goto the outer loop

return
; Retrun from the *Delay* routine

end

; must be put at the end of the program

;**

This is the end of the program. Now that you understand this basic
program of LED flasher, you can understand more complex programs and
tricks.

Let's speed up the rate!!

Now that we learned to flash an LED using the PIC 16F84, we'll increase the rate of information flow.

We'll write the same program using C language. This will be very easy as it is a popular language and it also a high-level language. Most of us used it for computer programming.

That is the flasher program:

```
//**********************************************

#include"pic.h"

main()

{

unsigned char i;

TRISB = 0 ; // Make PORTB output

PORTB = 0 ; // Initialize PORTB

for(;;)          // This is the infinite loop that keeps the PIC running

{

PORTB = 0x00;        // turn all LEDS off

for(i = 100 ; --i ;);      // Delay

PORTB = 0xFF;        // turn all LEDS on
```

```
for(i = 100 ; --i ;);      // Delay

}

}
```

//***

I guess the program is self explaining.

Now again, we'll try it in *Proteus* to see it working.

Here are the steps to add the C source code file and compile it.

1. Copy and paste the previous code into a file named *flasher.c* .

2. Install the program HiTec PIC C.

3. In ISIS 7, open the *flasher.DSN* model you used before.

4. On the Source menu, choose Define Code Generation Tools.

5. Press New.

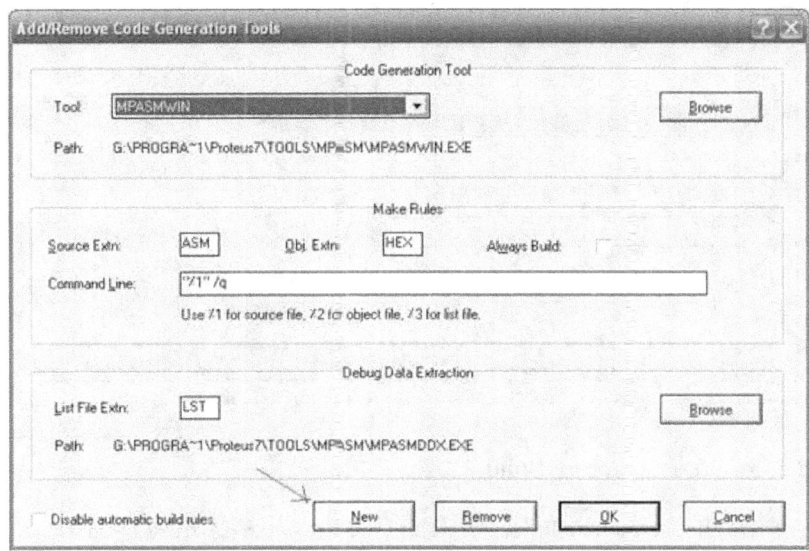

6. Browse to the folder of HiTec PIC C and to Bin folder and choose picl.exe.

7. To add the source , choose Source → Add/Remove Source files , press New to add

flasher.c choose PICC from the drop-down menu under Code Generation Tool , and on the

Flags text box , type --chip=15f84a to choose the microcontroller type.

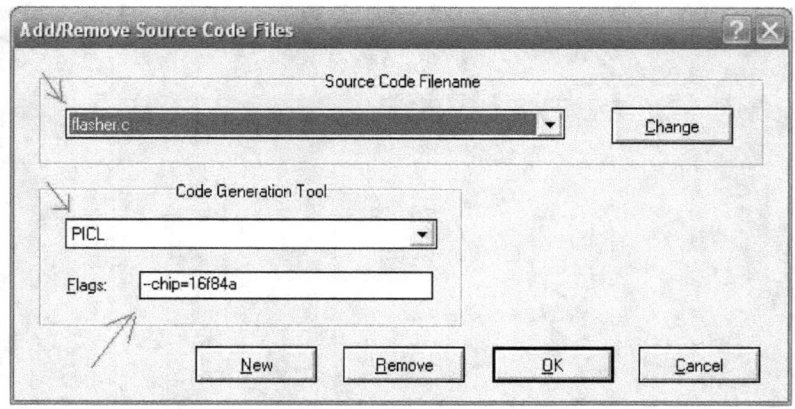

8. Now press Source → Build All.

9. Now choose the oscillator speed and the *flasher.HEX* file as you did in the Assembly

example.

10. Press F12 to Run the program.

Here you can find the Proteus ISIS model and the source code

Building your first circuit application:

Enough simulators!!

You may want to try the real circuit to see the microcontroller in action right now. I encourage you to build the circuit and feel it working. This will make you happy and will be a motivation to learn more about microcontrollers and make useful circuits.

So what do you need to get started?

- I guess you have the programming circuit and you have loaded the chip with the HEX file. If not yet, please do it now.

All you need you do is to put the chip in its socket in the programmer and open the program ICProg.

You need to configure the program as follows:

1) Form Settings menu, select Hardware.

- Select JDM Programmer.

- Select the com port you use.

- Select Windows API for Interface menu.

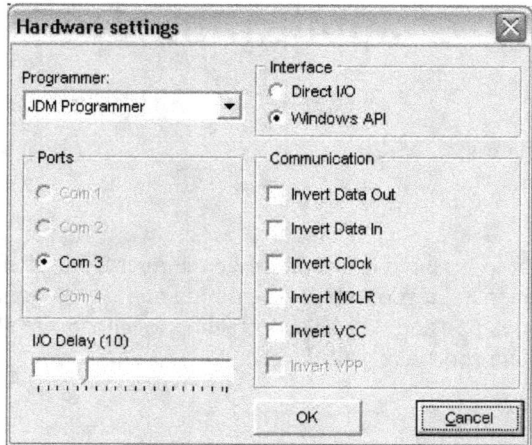

2) Select the microcontroller of your focus (Microchip PIC 16F84A) from the device menu.

3) Select the Flasher.Hex file to be loaded into the Microcontroller.

4) Choose HS for the Oscillator and clear WDT option.

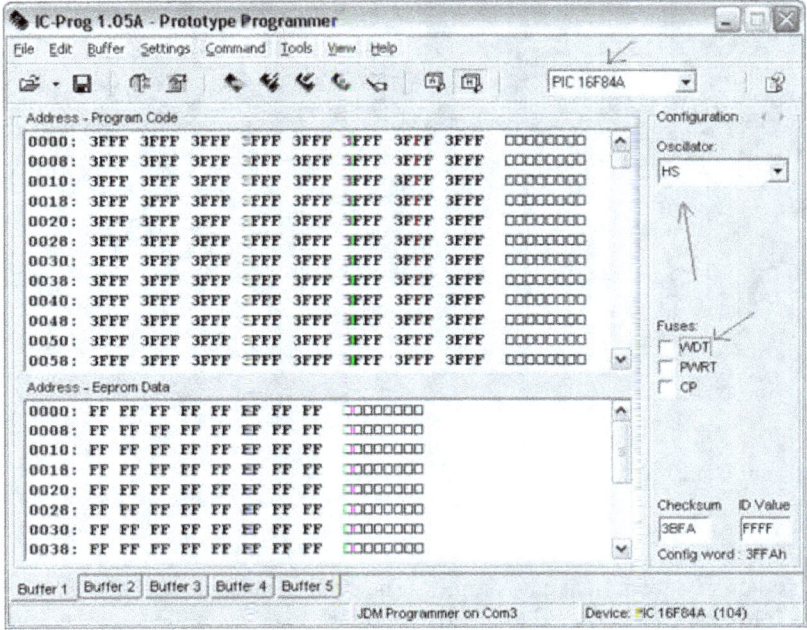

- Build the simplest application circuit for the PIC, the LED flasher.

Here is the schematic.

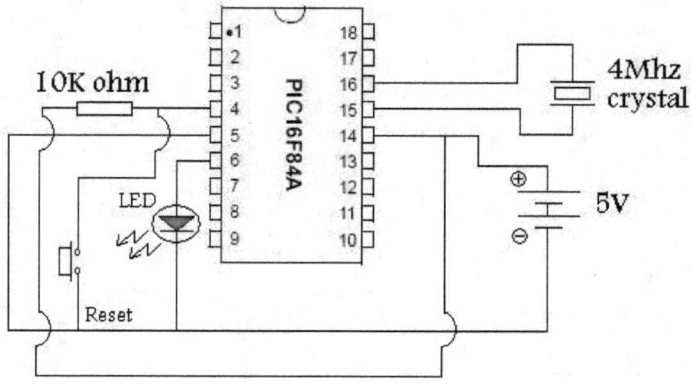

Simple, isn't it?

You don't need to make it complex. The push button connected to the pin 4 (~MCLR) is connected to demonstrate how to connect the circuit if you want to reset the microcontroller. If you don't need to do this (in this small project) connect the pin 4 to Vcc directly (+ terminal of the battery). This connection makes the microcontroller awake (not reset).

If every thing goes right, you get the LED flashing at a visible rate. Congratulations, you've just built your first embedded project!

Input to the PIC 16f84

Now that we learned how to get output form the PIC *(flashing LED)*, we'll learn to input the Microcontroller by a push button. This will give you an idea to get order from the user to do an action *(open door, turn-off lights ...)*.

```
; Button.ASM
;*********************************************************

list p=16f84

include "p16f84.inc"

org 0x00

goto start

org 0x20

start

bcf INTCON,7

movlw 0x00

bsf STATUS,5

movwf TRISA

bcf STATUS,5

movlw 0xFF

bsf STATUS,5

movwf TRISB

bcf STATUS,5

again
```

```
btfss PORTB,0

call LED_ON

call LED_OFF

goto again

LED_ON

movlw 0xFF

movwf PORTA

goto again

LED_OFF

movlw 0x00

movwf PORTA

goto again

end
```

;**

The program makes the LED is ON when the Button is pressed, LED off when Button is released.

The only new command here is BTFSS which checks if the button is pressed or released.

What it makes is Bit Test F Skip if Set where F is a bit in register. This command is a bit-oriented command. It means it deals with one bit of the register.

Build the following circuit in Proteus 7 ISIS:

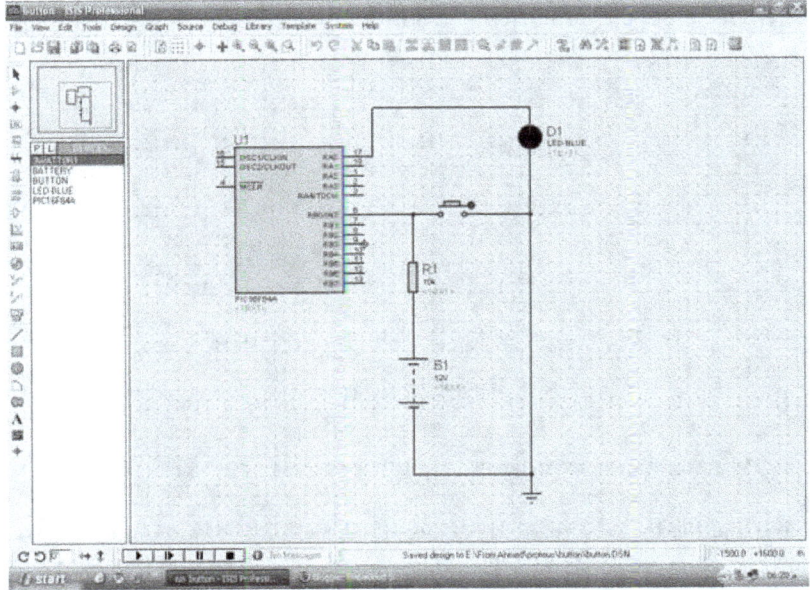

Follow the steps you learned before to add the source code to the circuit and configure it, and then start simulation. Of course in this time you'll add the *Button.ASM* source code.

Now, we'll write the same program in C

//***

#include"pic.h"

main()

{

unsigned char i;

TRISA = 0 ; // Make PORTB output

TRISB = 0xFF ; // Make PORTB input

PORTA = 0; // Initialize PORTB

```
for(;;) // This is the infinite loop that keeps the PIC running

    {
if ( PORTB == 0 )

PORTA = 1;

else

PORTA = 0;

    }

}
```

//**

Note that the logic is inverted because the button by default inputs *one* to the PIC and when pressed inputs *zero*.

Again, you can configure the project for the C source code from this lesson.

You can download the project files and source code from *here*.

Enjoy.

Introduction to Interrupt

What is the Interrupt?

What's the difference between polling and Interrupt?

Let 's take an example when you need to know if you have a phone call , you can know this using one of two ways :

1. Go to the telephone and pickup the handset to see if you have a phone call. You'll need to do this many times. And you will stop what you do every time.

2. Wait until the phone rings, then you go to it and pick up the handset.

Polling is like the first example and Interrupt is like the second example. You need to use interrupt in some applications. It sure decreases the load on the microcontroller and saves its time while doing another thing.

There are two types of interrupt: software and hardware interrupts.

In the following application, we will use Timer interrupt (one of the Software interrupts) which is initiated by the Timer0 overflow.

The timer is a free counting register that increments depending on the crystal oscillator speed and the **prescaler** settings. We can use the timer to calculate accurate time periods. And because we are using interrupt, we don't have to keep track of time while we are doing another thing. In the following example, the main program functions to monitor a push button and displays a number on a 7 segment display. The number on the display indicates the period in seconds.

Without using interrupt, we would have to calculate the time of the period and taking the display time into account.

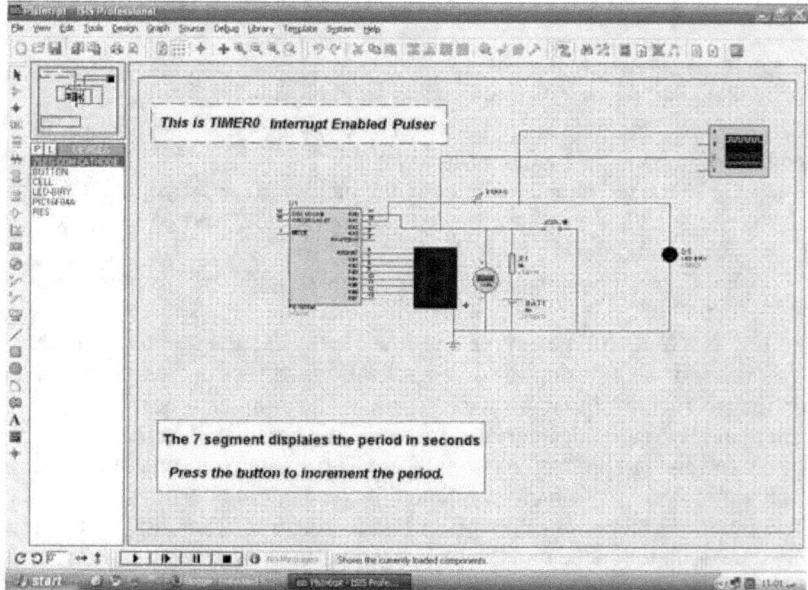

Here is the link for the project file and code.
Today, I revisited this page which contains many Microcontroller projects. This is the Cornell University Electrical Engineering students projects page. Many ideas can be found here.

Lookup table Sine Wave Generation

Sine Wave Generation using PIC 16f84A:

In this lesson, we'll learn to use the PIC 16F84A to generate a sine wave signal using lookup table method.

These are pre-calculated values of a certain signal (here *will be a sine wave*), then the signal is transmitted at execution time. You can consider this an inverse process of the digitization. At execution time, the signal is sent to the output port of the Microcontroller at the rate you determine. You then get the original signal.

Attached a text file that explains how I calculated the values.

The hardware circuit is very simple. It is called Resistor Ladder Network.

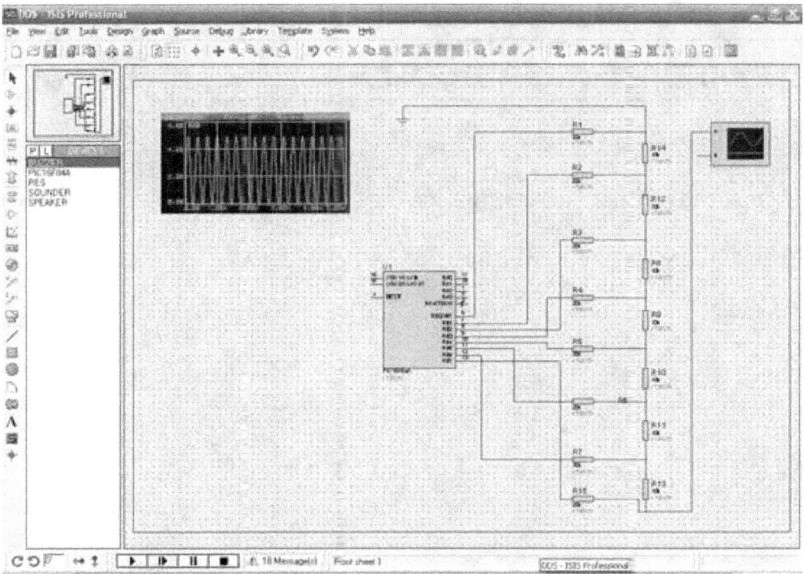

When you configure and run your program, you should get an output signal like this

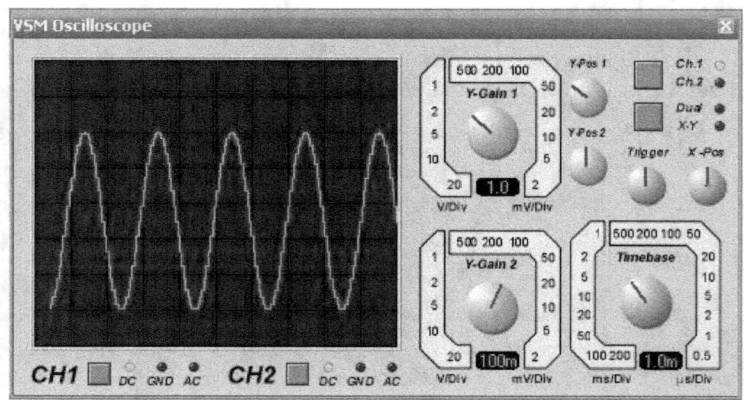

You can get the circuit for Proteus ISIS 7 and the code from here.

If you need any help for configuring the simulator, check my older posts.

You can also contact me directly if you need any further help.

PWM Mood Light using PIC 16F84

Generating PWM signal using PIC 16F84

PWM (Pulse Width Modulation) is a square signal involves changes in the duration of the on time (duty cycle). It can be easily generated by the PIC 16F84. This PWM signal can be used in many applications (for example: DC motor speed, LED brightness intensity, Mood Light Lamps).

PWM can be used when no DAC (Digital-to-Analog Converter) is available in the microcontroller chip or as a stand alone chip. The average DC voltage values are equivalent to analog values as where you used DAC for motor speed control.

And for light brightness intensity , at high frequency , the blinking process is not observed by human eyes and what you see is only brightness intensity corresponding to the duty cycle (time at which light is ON)

At longer duty cycle, LED is ON for longer time, so you see it brighter.

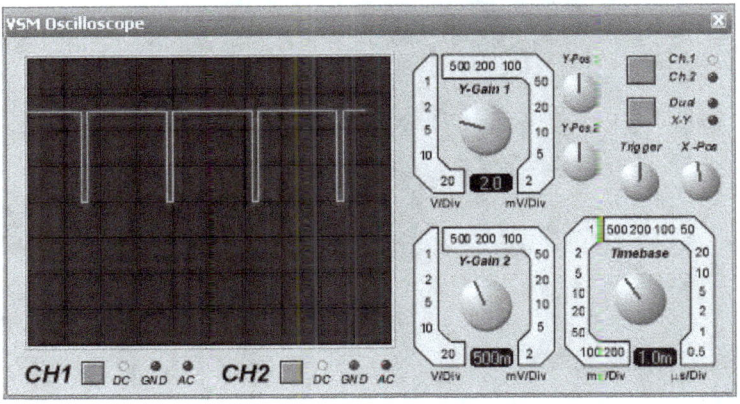

At shorter duty cycle, LED is ON for shorter time, so you see it less bright.

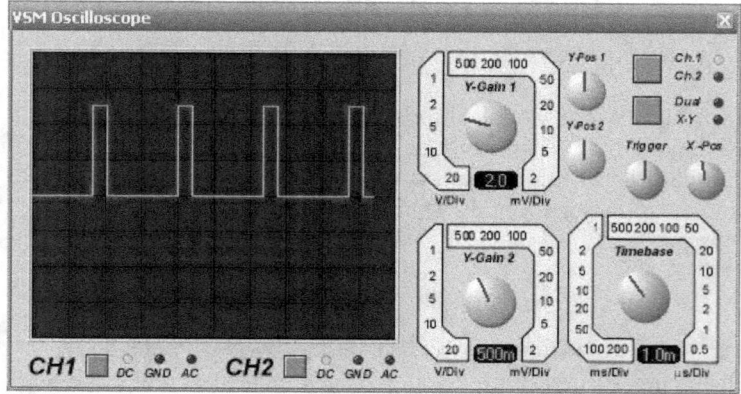

The previous examples used constant duty cycles at each time which means constant brightness LED and constant speed for motor.

To make Mood Light LED, you need to change the duty cycle of the PWM signal by increasing and decreasing it back and forth.

You get this one.

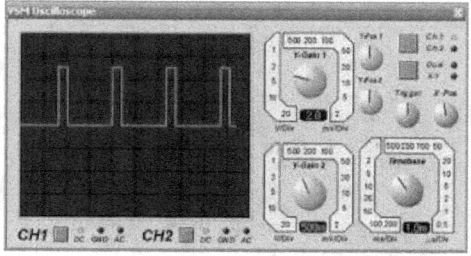

Note:

You see this image still, just click it and save it to your PC you'll see it animating (PWM is variable).

I connected this circuit to the LED and it did just as I expected, it looked brighter then dimmed and then brighter again.

You can have the circuit and the code files from here.

If you need any help configuring the compilation settings for Hi-Tech and ISIS 7, you can check the previous posts. If you need further help about code and hardware you are welcome to contact me.

DC MOTOR PWM CONTROL

In this post, we'll learn the DC motor control using PWM signal. We learned how to generate PWM signal using PIC 16F84 Microcontroller.

Hardware:

The PWM signal drives a 2N2222 transistor which acts as an electronic switch. The transistor switches the motor driving current on and off at high rate.

In the following figure, there are two circuits. The control circuit (the microcontroller) and the driving circuit (the motor and the transistor).

The DC motor draws relatively high current than the current in the microcontroller circuit. The transistor isolates the two circuits.

By changing the duty-cycle of the PWM signal, we get different average DC voltage for each duty-cycle value. The result is the change of the motor speed corresponding to that duty-cycle.

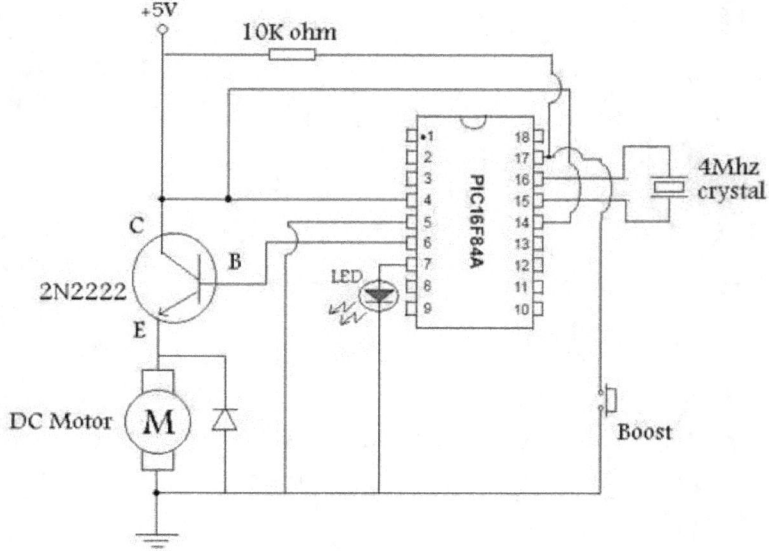

Software:

We use the PWM routine with two defined duty cycle values. The duty cycle value *duty1* is the larger value which drives the motor faster than *duty2* value. When the Boost button is pressed, the first duty cycle value is selected. When the button is released, the second value is selected.

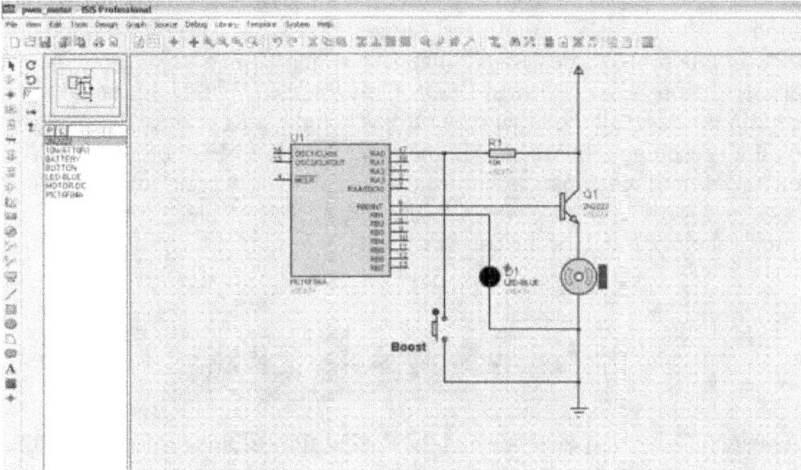

The software routine is very simple. As usual, you can download the Proteus 7 model from here.

If you need any help or have any feedback, just let me know.

LED Chaser: Larson Scanner

I love LEDs!!!

In this lesson we'll not learn any new technique from the technical point of view. But this is a cool project. I love LEDs. That's true. I cannot deny that. In this project I designed a nice and a simple thing. It is called a chaser. It consists of a row of LEDs. Only one LED is ON during a certain time and all the others are OFF. Then the next one is ON. And so on until reaching the last LED in the row. Then the LED before the last on is ON until reaching the start of the row. The light continues going back and forth. For those who saw the series 'Night Rider', it reminds you of the front logo of the car *Kitt*.

As usual, I give you the code and the design files of this project to see the simulation running on the Proteus ISIS 7 *here*. And for those who like it, the project can easily be built within minutes.

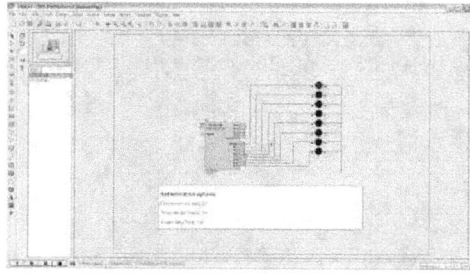

If you need any information, you can comment or send me an Email.

Thank you for dropping by.

PIC 16F84 Sound Generation

Do you know that you can generate tones of sound with the PIC16F84?

Yes you can generate these tones similar to the Midi tones of the old mobile phones. This is a very simple and funny circuit and you can try it on the simulation software Proteus ISIS 7 before building it.

Actually the circuit is so small it nearly contains no components other than the microcontroller, the crystal oscillator and some resistors.

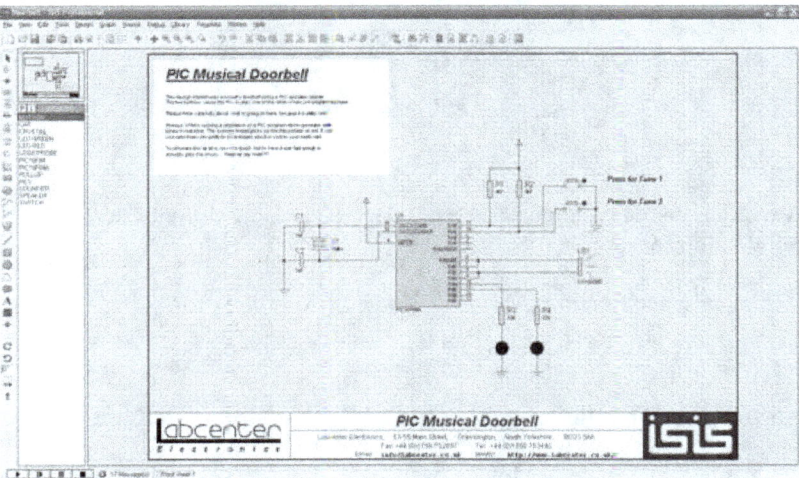

The software generates the tones according to the predefined words which represent tones.

The circuit lets you choose one of two tones with two push buttons.

You can find the Proteus ISIS Design files DSN, Code (Assembly) and HEX file as a sample project in Proteus ISIS 7. From the menu bar, press open file. You'll find the project under the Proteus → Samples → VSM for PIC16 → PIC Doorbell.

So, open the design now and if you like, start building the circuit and make you own tones.

Moway robot

Moway is a robotic system based on the famous Microchip PIC Microcontroller. It is very useful for students of robotics and embedded systems courses.

Moway is a small autonomous robot designed mainly for practical applications of mobile robotics.

Technical Specifications:

The robot is controller by a 4 MHz PIC16F876A Microcontroller.

As for the drive system, Moway has a dual servo-motor unit enables it to move.

The drive system is a closed loop system that uses PWM signals to control the speed of the motor and an encoder signal to measure the speed.

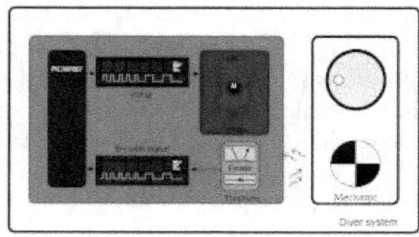

Motor control

The movement of Moway can be controlled through different parameters:

Speed control, Time control, Distance traveled and Angle control.

Also, Moway is equipped with sensors used for advanced autonomous motion control:

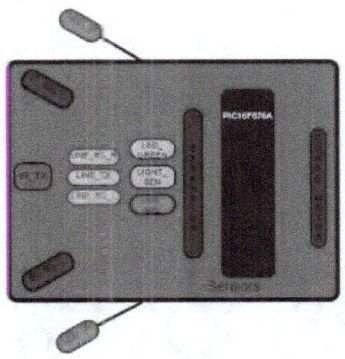

Sensors and Indicators

Line sensor: Mounted on the lower front part of the robot to determine the shade of color on which the robot is standing.

Obstacle detection sensors: These sensors operate in the same way as the line sensors, but they can determine the presence of obstacles (in digital mode) and the distance of it (in analog mode).

Light sensors: Measures the intensity of ambient light and the change in it.

Moway has an expansion connector to which an RF module can be connected to communicate with other Moways or with a PC.

There are four LEDs used for indication.

http://www.moway-robot.com/

PIC 16F84 Tetris Video Game

This is an old project that uses the Microcontroller chip PIC16F84 for the famous game Tetris. The output of this circuit is displayed on a TV screen. It also produces sound signals and can be controller by joystick.

I didn't design this circuit nor wrote the software for it , but I found it here and assembled it and found it very interesting. That's why I wanted to share it with my friends.

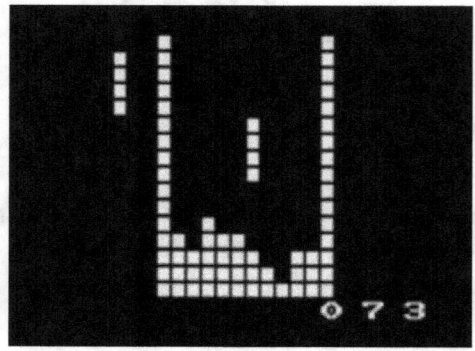

Here is a screen shot of the game.

And this is the schematic diagram for the circuit.

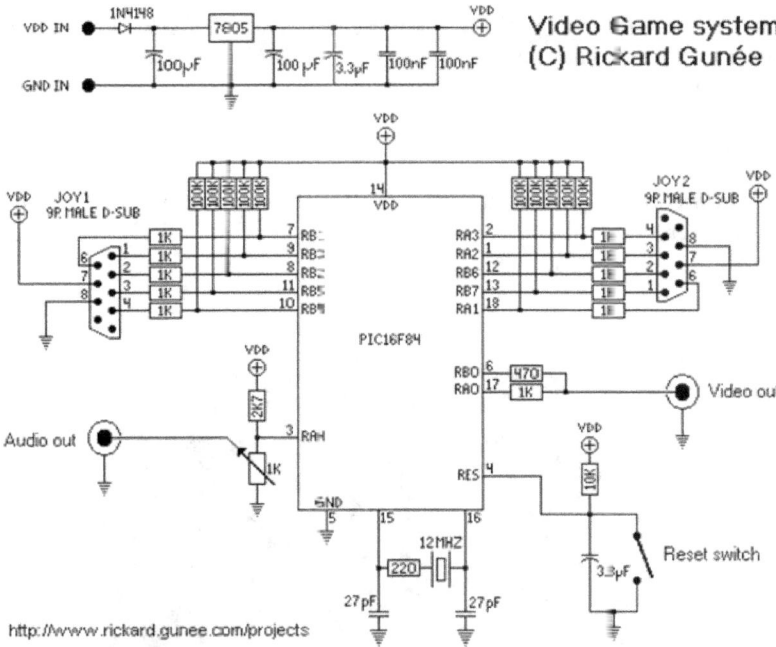

There is a free Emulator for the Microchip PIC that contains TV screen plug-in for viewing PIC signals that is directed to the TV. I found it here. I also uploaded the file here.

You can get the project package form here . It contains the schematics, PCB and software.

16F84 POV AirText

POV (Persistence of Vision)

This is a post of LEDs project. The project is called POV (Persistence Of Vision). If you are not familiar with this concept, it works as an LED matrix display.

The difference is that POV display consists of one column of LEDs only and is mechanically scanned through space to give the vision of 2D LED Matrix Display.

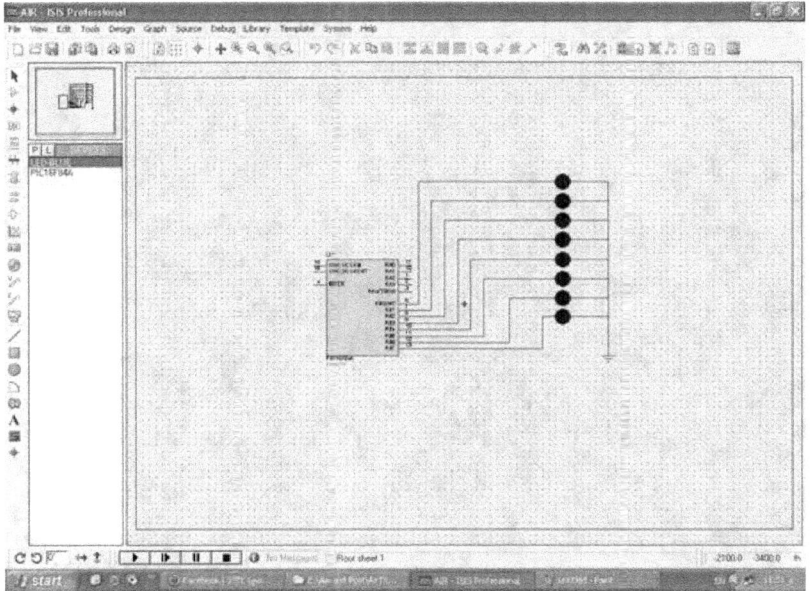

Many projects are built in rotating (circular motion) or in oscillating layout (moving left and right).

I've seen this project "AirText" based on the Sun Microsystems Sun Spot, and then I wanted to make a similar yet more simple project.

There are many Microcontroller projects based on the POV concept. But i think this is the simplest circuit that can do the job with minimum hardware.

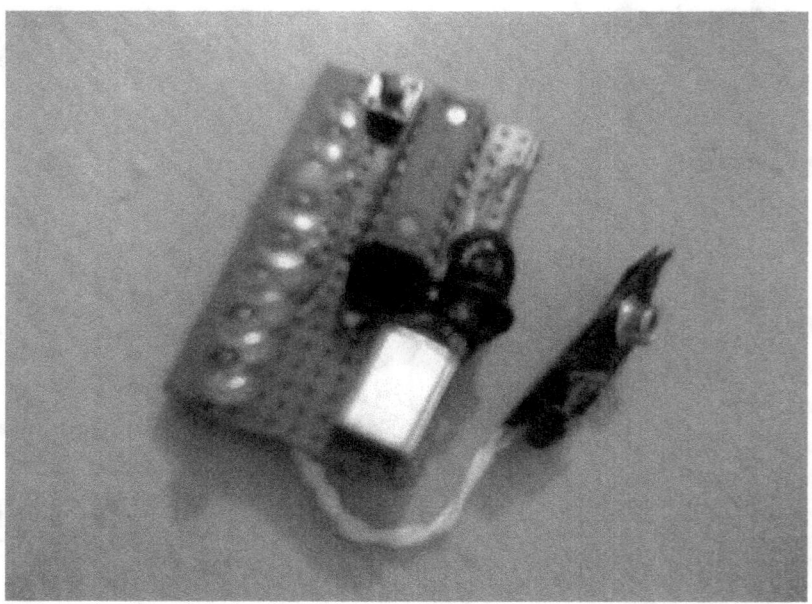

The only drawback with this circuit is that it has no sensor for direction of motion. The

Sun Microsystems Sun Spot has a built in motion sensor called accelerometer. This senses the amplitude and direction of acceleration. So when the module is moved in one direction it draws the scanned characters according to this direction.

But with this simple circuit using PIC16F84, the characters are drawn in one direction independently of the circuit direction of motion.

You can find the source code, Hex file and Proteus 7 des gn file here.

You can read this post in Arabic

http://arabic-embedded-egypt.blogspot.com/2010/05/pov-persistence-of-vision.html

Here is the project on my profile at instructables.com

http://www.instructables.com/id/Cheap-16F84-POV-Message-AirText/

LCD Interface to PIC 16F84

I really love this Microcontroller PIC 16F84. I am always searching for programs and projects and piece of code written for this old Microcontroller.

Today I found code on the web for this chip interfacing it to a standard LCD module.

http://www.electronic-engineering.ch/microchip/projects/LCDx_test/LCDx_test.html#assembler_code

It is written in assembly language and I am not going to explain it. So I'll leave it to you to understand.

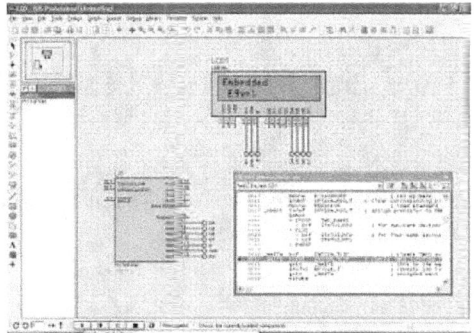

I only made a Proteus 7 model for the circuit that you can find it here. I promise you to make a new post for the LCD interface using the embedded C language.

7 Segments POV

This is a post about the POV (Persistence of Vision) Display that uses one 7 Segment display.

In short, I made this little project as a proof of concept of something I've always noticed in many electronic devices those used 7 segment display.

In those devices, the 7 segment display was flickering instead of lighting it with steady volt and displaying the number accordingly. That flickering effect may be nice to some people and may be annoying for others.

So I decided to build this project so fast on a breadboard and wrote the C code for it.

In this image you can see the 7 segment connected to the PIC16F84A Microcontroller on a breadboard and flickering with numbers.

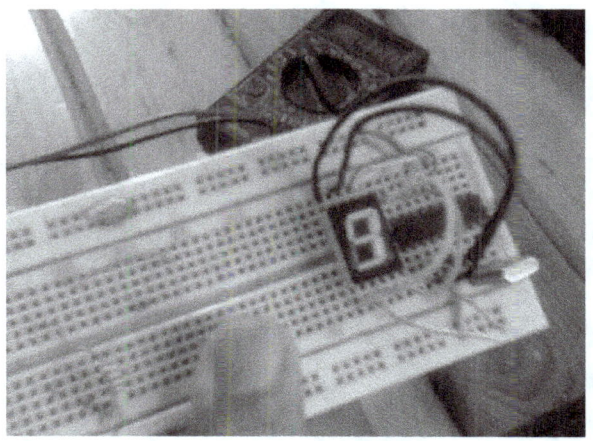

I just made this 7 segment display flickers with different numbers

I thought that if those flickering devices displayed many numbers at high speed and have been moved, guess what, the POV effect will be produced.

That's right. The same POV effect we've seen made using LEDs can be produced by a single moving 7 segment display.

Here, you can see some pictures of the breadboard while being waved to the left and right in front of the camera.

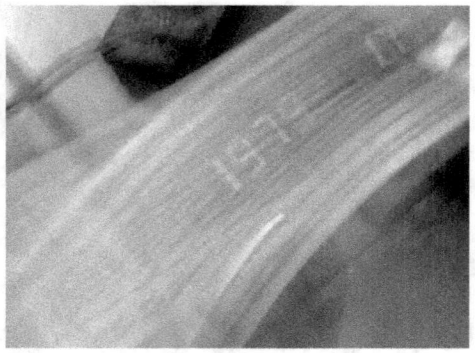

The POV effect here makes our eyes (and the camera too) see the numbers as separate from there each other as four digits.

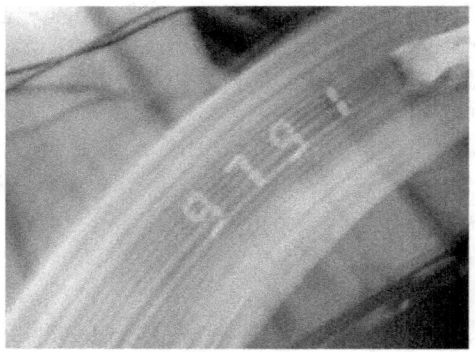

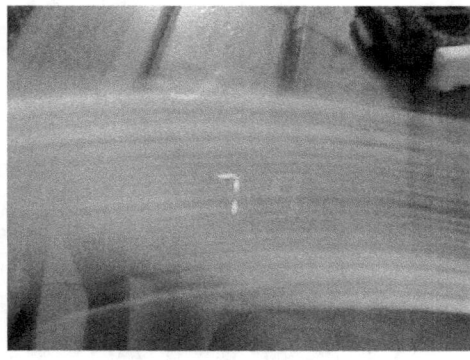

The circuit is very simple. I used a common anode 7 segment display directly driven by the PIC16F84A. And written the program in C code. The 7 segment patterns are directly sent to PORTB to be displayed on the 7 segment display.

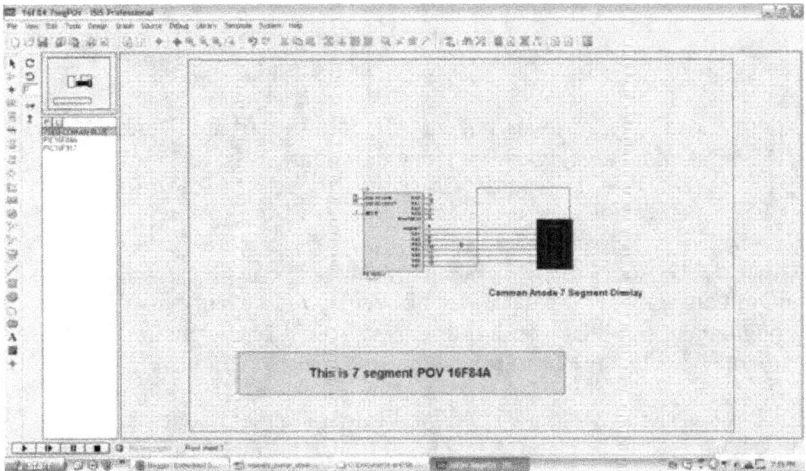

You can find the Proteus 7 model and C program here.

I hope you enjoy this project.

Hacking Infrared with PIC 16F84A

Send an Infrared signal with PIC 16F84:

In this post, we'll learn about transmission of the Infra red signals using PIC16F84. You can use Infra red signal in many projects (Robots, Touchless counters, Theft alarming ...)

All you need to know to start using Infra Red signals are IR transmitters (IR LEDs) and IR receivers (IR receiver modules).

The IR LEDs are normal LEDs in the outer shape but they emit IR signals and of course these signals cannot be seen by the human eye. You can though make sure it is working fine by using a digital camera. The digital camera can detect the Infra Red signal and makes them visible to you.

This image shows you how the digital camera in the mobile phone can make you see the IR signal transmitted from a normal IR remote control.

The other difference between visible light LED and IR LED is that in visible light LED, the +ve terminal (Anode) is the longer one. But in IR LED, the +ve terminal (Anode) is the shorter one.

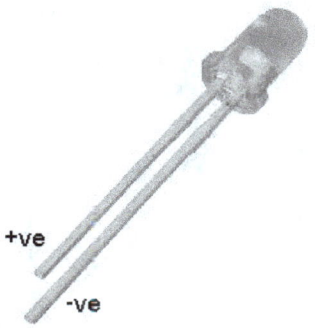

+ve

-ve

The second thing you need to know is the IR receiver module. It is an IR receiver component connected to a demodulator and amp ifier circuit. The function of the demodulation circuit is to demodulate the modulated signal received by the receiver module. This means that this module can receive and understand ONLY modulated signals!!!

YES, that's true. And this is used to differentiate between the IR communication and control signals and IR signal emitted from all the objects around us and from all visible light sources. The demodulator circuit has a Band Pass Filter (BPF) that can detect only signals modulated by a 34 kHz carrier signal. Here is a datasheet for TSOP312 Infrared receiver module .

1
2
3

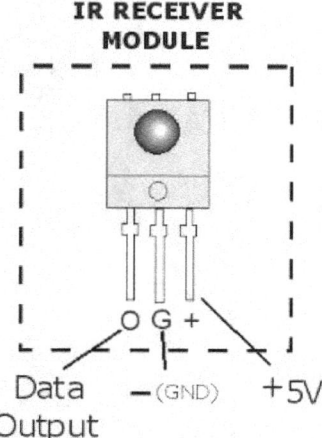

IR RECEIVER
MODULE

Data   —(GND)   +5V
Output

This means that if you want to transmit a signal that can be demodulated as (1), you need to send a continuous signal of 34 kHz from your IR transmitter LED.

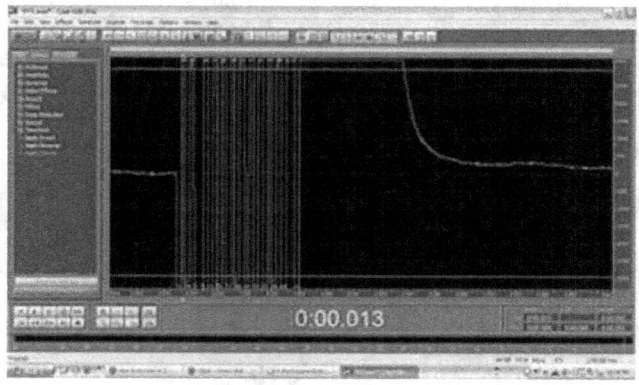

0:00.013

This is the signal I captured from the Remote Control by the IR receiver

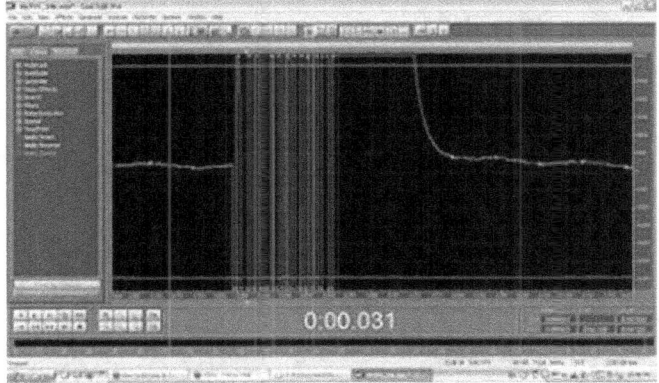

And this is the signal I generated from the PIC 16F84A

The software is very simple this time and written in Assembly Language. All it does it sending pulses to the IR LED at the correct frequency for the correct periods of time. It consists of a 34 KHz carrier modulated by the signal of consecutive zeros and ones.

I knew the time coding of the TV remote control by capturing the signal using the IR receiver module to the sound card of the PC. Then I viewed it using sound editing software to show the exact times for the IR signal.

Here is the link for the Proteus 7 model and software for the project.

16F84A VGA Output

Have you ever thought of this?

That you can generate any kind of signals using your little Microcontroller PIC 16F84A. And one of these signals is the VGA signal that appears on your PC monitor. I thought of this idea but I couldn't implement it. But I found it at this website. So I wanted to try if it is working and share it with you.

I found an old web page containing this project http://tinyvga.com/pic-vga . I just made some minor modification on that project and put it into a real circuit. I found that the there is a small timing issue in this software that makes the monitor flicker repeatedly.

The circuit is very simple. It consists of the Microcontroller PIC 16F84A and the 15 pin female VGA connector.

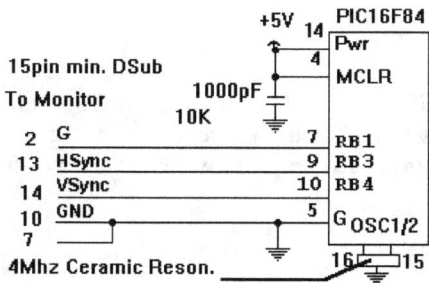

This is the schematic diagram of the circuit

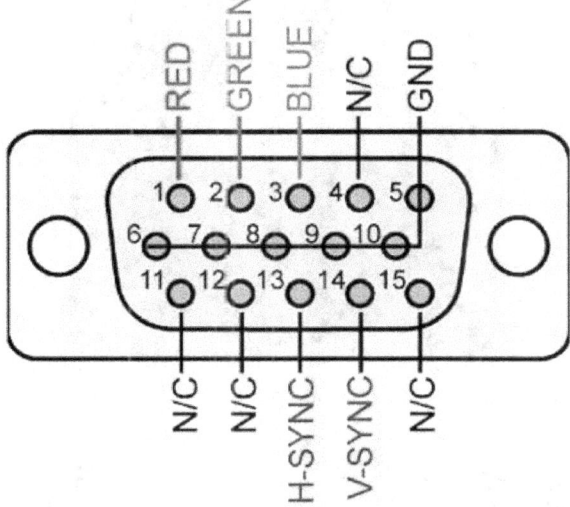

This is the pinout of the VGA connector

There is a simple hint for your when downloading the software on the Microcontroller. Make sure that the configuration is:

PRTE ON

XT

WDT OFF

I took this picture for the monitor after running the program.

This picture shows the circuit.

I'll try to fix this issue and post it when I am done with it. Now I will show you the results of the current software. I hope you enjoy it.

Here is the link for the Proteus 7 Model, source code and Hex file.

You can read this post on my page on instructables.com

http://www.instructables.com/id/16F84A-VGA-Test/

16F84 SERIAL COMMUNICATIONS

In many projects exists the need to exchange information with outside world. This case can be the need to store information from your project to a database on a standard PC or a need to communicate with another device. RS232 is widely used serial communications protocol. Modern Microcontroller chips contain hardware modules for such communications protocol. In those chips all you have to do is configuring the RS232 module and to choose the right crystal to produce the desired baud rate.

But in Microcontroller as 16F84A there is no such hardware communications module. Serial communication can be established by software or Bit-Bang method. The standard RS232 protocol uses +12v to represent logic 0 and -12v to represent logic 1. While TTL uses +5v to represent logic 1 and 0 V to represent logic 0.

An external IC is used as an interface between the Microcontroller and an RS232 enabled device (PC for example) . This IC MAX232 or HIN232 performs this level shifting function between the RS232 and TTL protocols.

Recently, I found a website implementing RS232 protocol in software using C language. The new method represented in this website is using the inverted logic (0v for logic 1 and +5v for logic 0).

This method enables connection of the Microcontroller directly to the PC. The web page clearly explains this method. The PIC 16f8A can withstand +12v and -12v when receiving data from PC. And for data transmission to PC 0v to +5v can be distinct by the PC as two different levels.

I really wanted to try this code in real world and built the simple circuit as shown in the web page. The designer recommends connecting a resistor for PIC protection.

Of course, you can find that it is a very simple circuit that provides another new usage for the PIC 16F84A.

So how can you use the serial data form the Microcontroller?

You can write simple software for the PC that reads from the PC's serial port (com1 or com2). Or you can simply open the famous program **Hyper terminal** that comes with all Microsoft windows versions under the menu Programs -- Accessories -- Communications -- HyperTerminal. Then you need to configure the right Com port, Baud rate and Flow control. As shown here

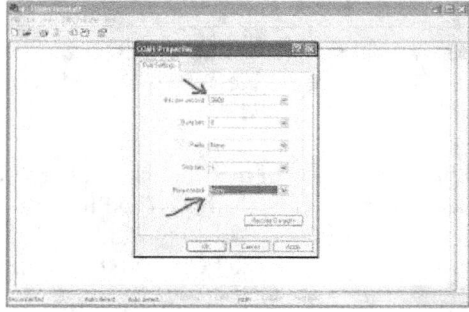

I used a serial connector from an old mouse to connect the circuit board to the PC

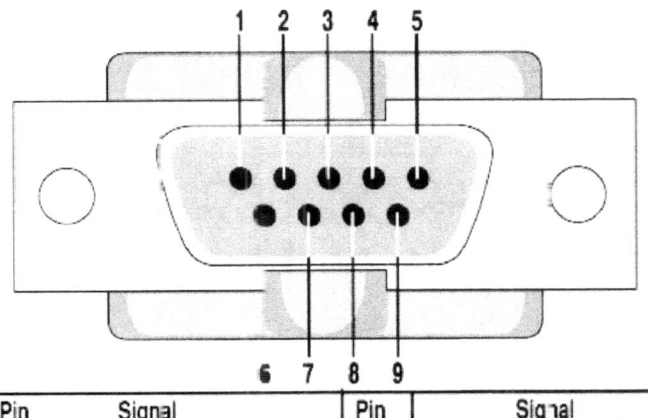

Pin	Signal	Pin	Signal
1	Data Carrier Detect	6	Data Set Ready
2	Received Data	7	Request to Send
3	Transmitted Data	8	Clear to Send
4	Data Terminal Ready	9	Ring Indicator
5	Signal Ground		

Here is the pinout of the 9 pin serial connector

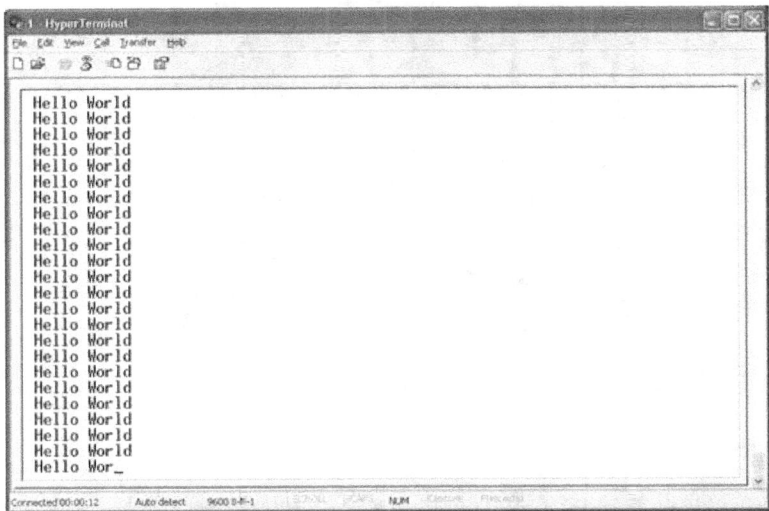

This is the Hyper terminal program showing "Hello World" message from the Microcontroller.

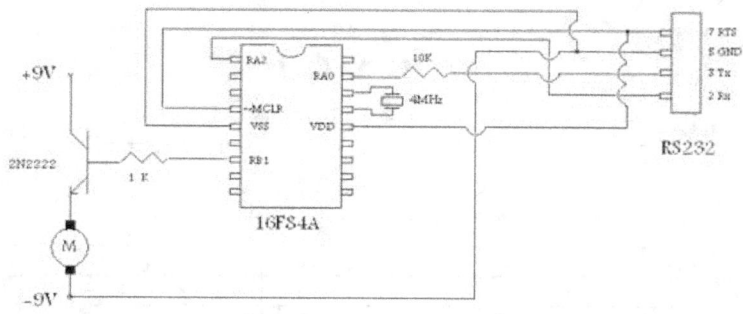

This is the schematic drawing of the circuit

Here is the real circuit board I built.

Here, we'll use the serial communication protocol RS232 to adjust the Microcontroller which in turn controls a DC motor by the PWM signals.

By choosing the motor speeds and sending them to the Microcontroller, we are able to choose PWM signals duty cycles and sending them to the Microcontroller via RS232 protocol.

The circuit between the Microcontroller and the PC is very simple. All you have to connect is the PIC16F84A, the crystal oscillator and the RS232 9 pin female connector.

The Microcontroller is supplied by the 12V DC from RS232 pin 7 RTS. Do not be afraid of losing the PIC chip, it can withstand it without damage.

The PIC16F84A TX pin 1 is connected directly to RS232 pin 2 Rx.

The PIC16F84a Rx pin 17 is connected to RS232 pin 3 TX through 10k ohm resistor to adjust the input voltage to the PIC.

The motor driving part of the circuit has already been explained in the PWM DC motor control post. So I'll use the same driving circuit.

The PWM signal comes from the PIC16F84A to the NPN 2N2222 transistor base. This transistor acts as an electronic switch.

The DC motor is driven by an external DC 9v battery.

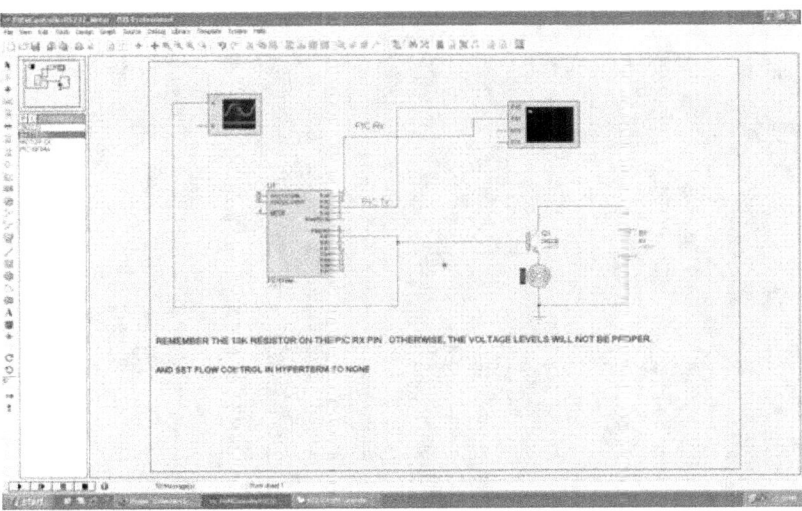

This image shows the Proteus 7 Model of the project

As usual, you may want to try the code and the circuit on the simulator first before building it. So, you'll find the Proteus 7 model and code in this link. But to properly send and receive data form PIC 16F84A to the virtual terminal model in Proteus 7 environment it must be configured to use inverted data. You'll find that I configured this for the model. But if you want to see it for yourself, right click on the virtual terminal model and choose inverted.

You can read this post in Arabic Language

http://arabic-embedded-egypt.blogspot.com.eg/2010/09/blog-post.html

16F84A Spindle Motor Control

In this post, we'll see how to control a spindle motor using the Microcontroller PIC 16F84A.

The spindle motor is a special type of Brushless DC (BLDC) Motor that runs on high speed.

Unlike other types of motors, spindle motors have more than 2 terminals (They can have 3, 4 or five terminals).

Thus, the spindle motor is driven by switching its coils ON and OFF according to certain sequence. The function of the microcontroller is generating the driving signals.

Of course, the output current of the Microcontroller is not sufficient to drive the motor coils. There is an interfacing IC that can supply the motor coils with the suitable driving current. This IC is ULN2003. It acts as an electronic switch opened and closed according to the output signals of the Microcontroller.

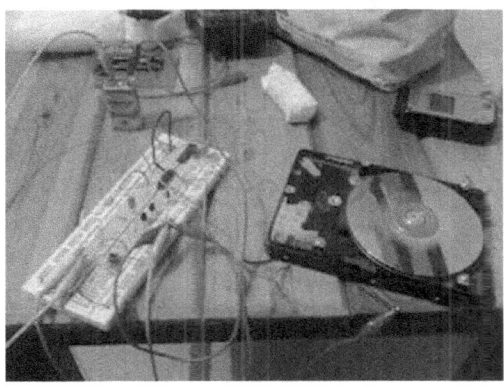

This is the real circuit connection

As you can see, I used an old hard disk drive motor. This is a three coil spindle motor. Actually it has 4 terminals (3 Coils and 1 Common).

You can see the four terminals on the back of the motor

You can find the source code and simulation files on this link.

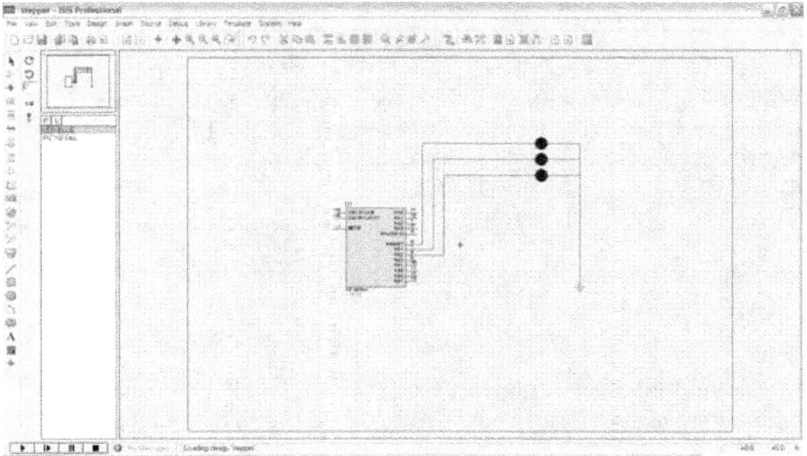

The simulation model shows LEDs connected to the Microcontroller 16F84A. That is because this version of Proteus ISIS does not contain a three coil spindle motor model.

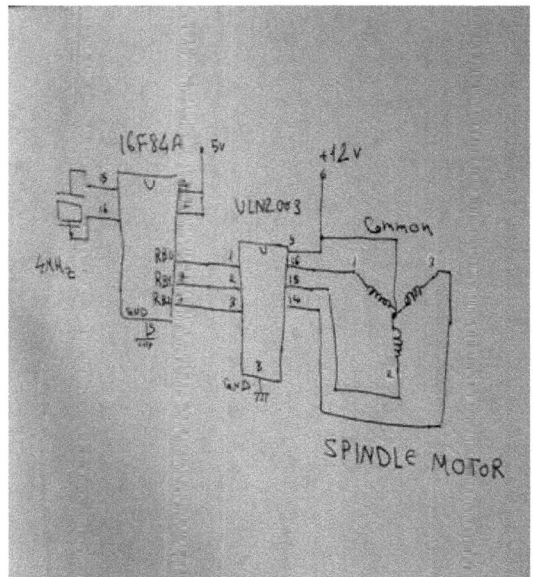

This is the schematic diagram of the circuit.

16F84A Frequency Counter

While I was searching the web for some Embedded Systems projects, I found a web page for a Frequency meter based on my favorite PIC16F84A.

I decided to build the Proteus Model for the circuit and test it.

I found the circuit interesting because I tried to build a frequency meter once before using a 7 Segment Display. I will post it in my blog later.

Here is the Proteus ISIS Model I built

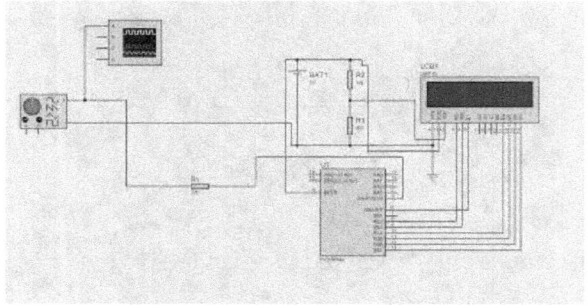

The circuit hardware is very simple.

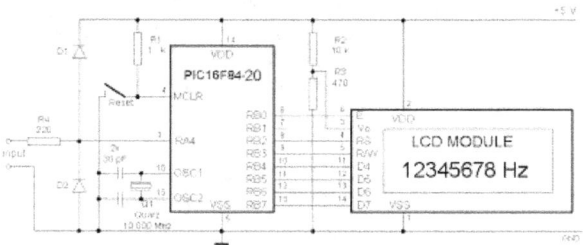

The software is written in Assembly Language.

Here is the link for original project.

http://blog.savel.org/2006/02/05/16f84-frequency-counter/

And here is the link for the Proteus ISIS Model

New Microcontroller Chip

Microchip PIC 16F917

I used the Microchip PIC 16F917 in simple programs and tried the good features of it. I ordered some samples from Microchip.com and they sent them to me. I searched online for a simple programmer that I use it to program the chip the same way I do with the PIC 16F84. I found it *here* . And I uploaded it *here* in case you couldn't find it.

The programming circuit is very simple and uses the serial port.

Here is the schematic diagram:

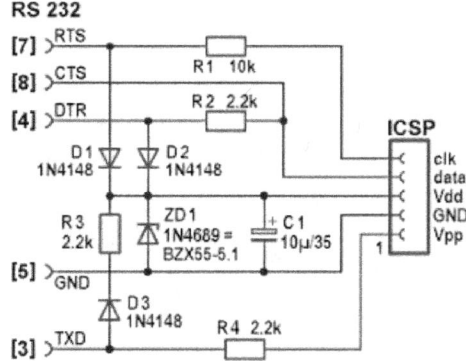

After you build this simple circuit, you can use the program to load the HEX file.

This is a snapshot of the software:

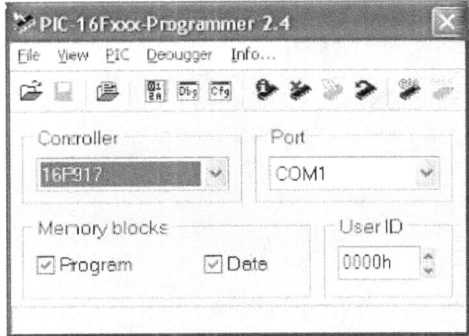

The Microchip PIC 16F917 microcontroller is a very nice and modern chip. You can find its datasheet on Microchip website.

PIC16F913/914/916/917/946
Data Sheet

28/40/44/64-Pin Flash-Based,
8-Bit CMOS Microcontrollers with

You can see the circuit I built lately here:

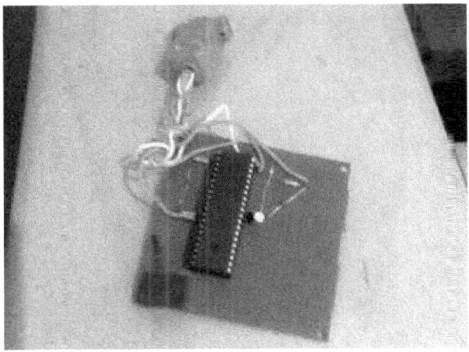

You'll see more projects using this chip. And I'll be posting projects for the PIC 16F84.

By the way, The PIC 16F917 is backward compatible with the PIC 16F84. That means if you write a code for PIC 16F84 in assembly or C or even compiled it to HEX file, you can still use it with the PIC 16F917 without a change in code or a recompilation.

Happy programming. Have a nice time.

16F917 CCP Block PWM - 16F917 PWM Generation

PWM can easily be generated on the PIC 16F917 using the CCP block. If you are new to this concept then let me introduce it to you.

New Microcontrollers contain dedicated hardware blocks for generating signals and for doing special functions. An example of these blocks is the CCP block in many Microchip Microcontrollers.

In this post, I'll introduce you with the CCP block of the PIC 16F917 Microcontroller.

It is called Capture, *Compare and PWM block*. Here, we explain the PWM function.

As the name implies, it generates PWM signals in hardware instead of the software PWM signal we generated using the PIC 16F84 in an old post of this blog.

The advantage of this concept is that you save code space and processing time for another functions and tasks.

So how can you configure this block?

You'll find an extensive explanation for this topic on the PIC 16F917 datasheet.

And I used this link to calculate the values for the block configuration registers.

http://www.micro-examples.com/public/microex-navig/doc/097-pwm-calculator

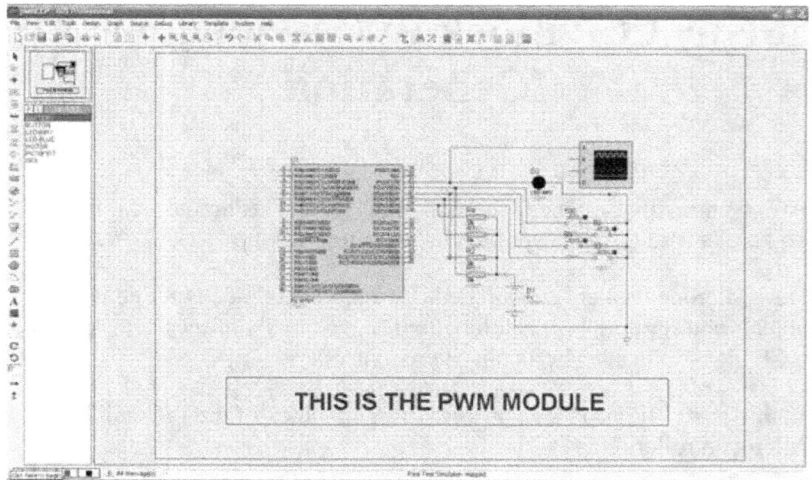

THIS IS THE PWM MODULE

You can find the design files and source code for the project here .

Now you can use this project to control any circuit with PWM signals generated with PIC 16F917 CCP PWM block as you did before by using PIC 16F84.

You can control DC motor speed, LED intensity and make Mood Light.

If you have any questions or feedback, contact me.

Yaw Rate Gyroscope to PIC16F917 (Analog-to-Digital converter Module)

In this post we will study the ADC (Analog-to-Digital) Module of the Microcontroller PIC16F917. We will study a real circuit of PIC16F917 interfacing to a semiconductor Gyroscope.

The Gyroscope is a motion sensor that senses tilt in a certain direction. The used sensor is a yaw rate sensor (free sample from Analog Devices. The Gyroscope ADXRS613 was sent in an evaluation package EVAL-ADXRS613). Yaw means rotation around the vertical access. And rate means the acceleration of this rotational motion.

This is Microchip PIC 16F917 Microcontroller

The output from this sensor is analog signal which represents the Yaw rate in certain direction (Left or Right).

The analog signals are converted in the ADC module of the PIC16F917 and are represented on a LED column according to the yaw rate and direction.

The program is very simple and straight forward. It starts by configuring the ports of the Microcontroller for input and output. Then the ADC

Module is also configured (Channel, sampling rate and result data format).

Then the infinite loop of the program starts which contains the step of

starting the conversion, waiting for conversion to complete and displaying the result on the LEDs.

When the circuit moves in clock wise direction, number of LEDs in one half of LEDs column illuminate according to the rate of change in angular motion.

And When the circuit moves in counter clock wise direction, the other half of the LEDs illuminate indicating change in direction and indicating rate of change in angular motion

ADXRS613 Evaluation Datasheet EVAL-ADXRS613

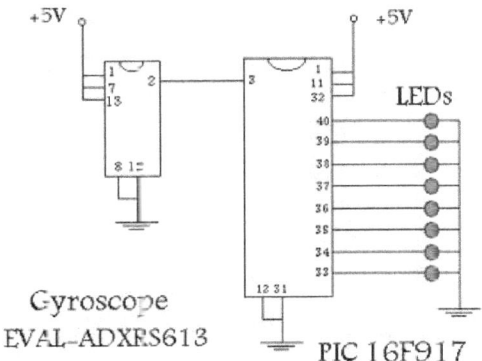

This is the circuit diagram

This is a picture of the real circuit

http://www.youtube.com/watch?v=LraBA6oD1jI

And this is a video of the circuit in motion

You can find the code and simulation files on this link

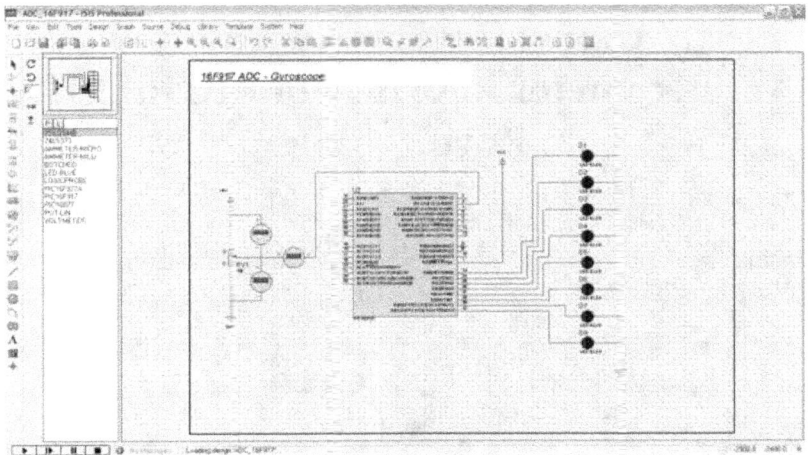

This is the circuit as shown on Proteus 7 simulation environment

The Gyroscope is replaced in simulation by a variable resistance because Proteus 7 does not contain a Gyroscope model.

Read this post in Arabic

Here is the article on my instructables.com profile

Analog Devices ADXL206 Accelerometer interface to PIC16F917 - Purple Cube

In this circuit we'll learn how to interface an Analog Accelerometer to Microchip PIC16F917

We use in this circuit the Analog Devices 2 Axis Analog Accelerometer ADXL206, shown above.

Here is the Proteus ISIS Model

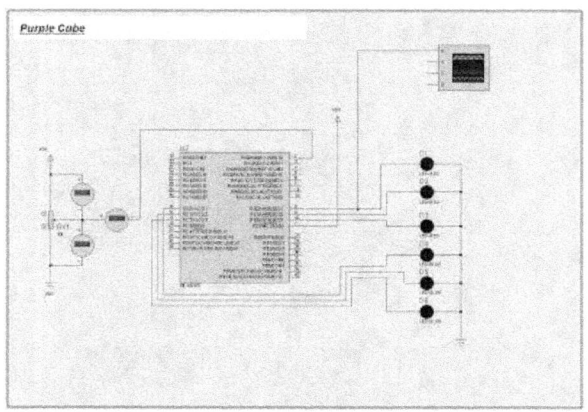

Hardware

Here are the circuit components

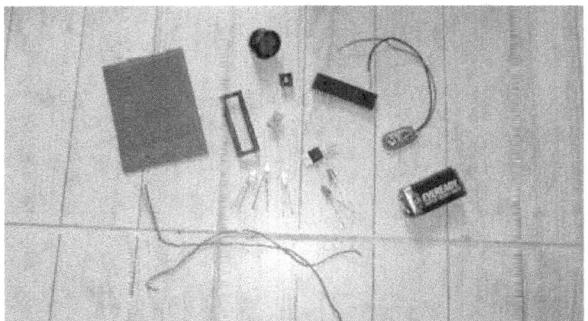

Here is the complete circuit diagram

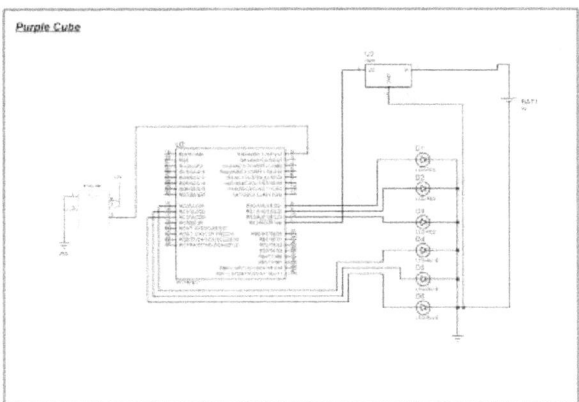

Here is the circuit after assembly

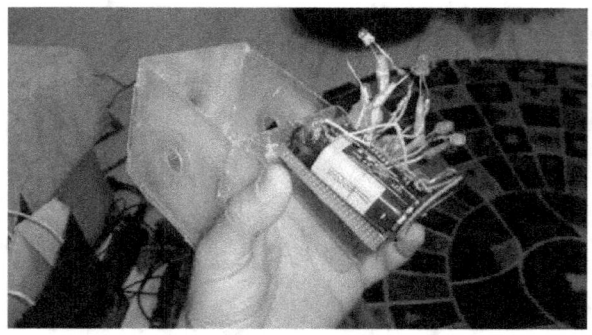

Software

I wrote the program in Embedded C Language and used HiTechPIC Compiler.

Here is the Proteus ISIS Model Files and Source Code

Here is the Full Project Steps on Instructables.com

PIC 18F4550 Programmer - The best is getting better

It was 2003 since I built my first JDM Microcontroller Programmer. Since that time, I didn't stop programming and building Embedded Systems based on the Microcontroller Microchip PIC 16F84A. I've learned many other Microcontrollers from Microchip and other manufacturers. But I still have the same love for this old Microcontroller and this beautiful programming circuit (JDM).

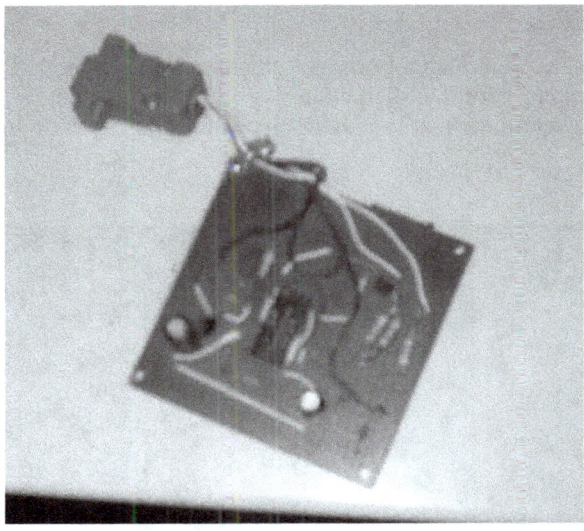

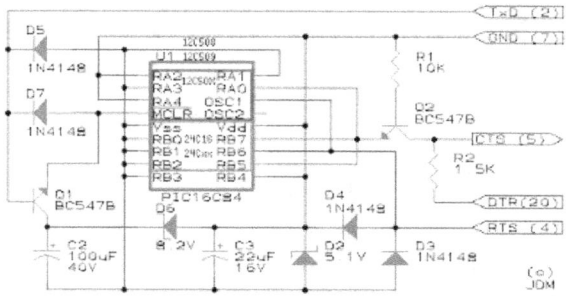

JDM original schematic

I built new Programmers for other Microcontrollers. Recently I searched for a programmer for the Microcontroller PIC18F2220 and I found a software programmer that uses JDM circuit for programming it.

And then I searched about the advanced Microcontroller PIC18F4550 which can be directly connected to the USB port. The best thing was it can also be programmed by the same JDM circuit. The only modification

I had to do with the circuit is adding a new socket for the new Microcontroller chips. The Microcontroller chip PIC18F4550 has 40 pins so I had to install an extension to my old JDM circuit because it had no space left.

That's the way I built this extension. A 40 Pin IC connector on small piece of plastic connected to wire by the conductive ink.

Basically, all the PIC Microcontrollers use the same pins for programming.

Five pins are used for basic serial programming:

Vpp   -----   Programming voltage

Vdd   -----   Vcc

Vss   -----   Gnd

Data   -----   Serial Data

Clock

This is the 40 pin IC connector on a piece of plastic

Then I welded the wires to the PIC18F2220 28 pin IC connector.

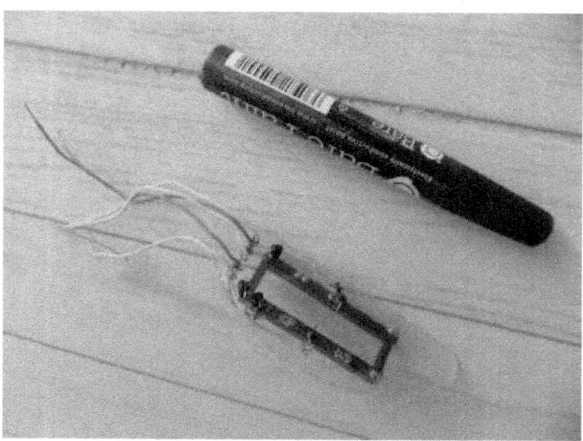

The conductive ink and the 40 pin IC connector welded to wires

This is the final circuit after assembly

Programmer software loader used with this circuit:

PicPGM   http://picpgm.picprojects.net/

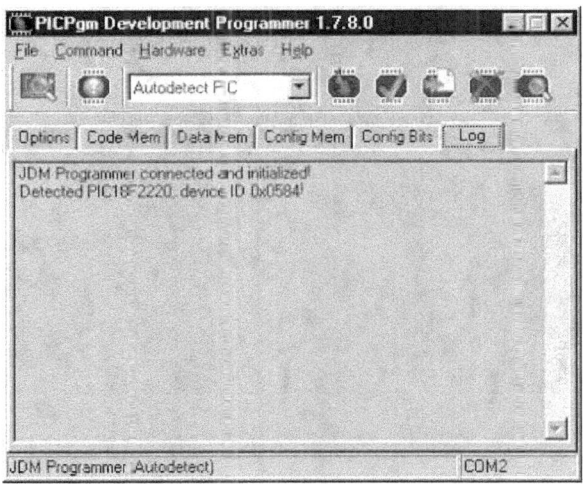

Screen shot of PicPGM

Read this post in Arabic

http://arabic-embedded-egypt.blogspot.com/2013/12/18f4550.html

This is my project post at instructables.com

http://www.instructables.com/id/18F4550-Programmer-in-Ten-years/

Pinguino Egypt - Do-it-Yourself PIC Arduino Clone

Pinguino Egypt - PIC Based Arduino

It's been eleven years since I first started learning and building my first Microchip PIC Microcontroller circuit. I really like this Microcontroller family.

Recently, I started reading and trying some Arduino.

I really wished to build a similar circuit based on the Microchip PIC. Even I started to think to design a circuit of my own. I searched online and I found Pinguino.

Just like Arduino, there are many versions of Pinguino using different Microcontrollers from Microchip. I wanted to build the version using PIC18F4550.

I found Pinguino before this time and saw it among many other Arduino clones, but this time was really different.

I was ready to start making circuits with the PIC18F4550 chip after I completed building the PIC18F4550 programmer.

So I decided to start from where the others have reached

I've decided to build this magnificent circuit that brings all my dreams together,

PIC based, Arduino compatible and Easy-to-build.

I was such excited to build it and try all Arduino projects with it. So, I started to collect the data and components to start working.

I also wanted to try the PC Software/Embedded Software communication.

Although I was determined to build the circuit on a copper board to stay with me and use it many times, I wanted to quickly assemble it on a bread board to try it ASP!!!

Here are some useful links:

http://wiki.pinguino.cc/index.php/PIC18F4550_Pinguino

http://wiki.pinguino.cc/index.php?title=File:Pin_pinguino_18F4550_%282%29.png&limit=500

http://blog.pinguino.cc/

http://www.hackinglab.org/pinguino/

http://jpmandon.blogspot.com/

https://sites.google.com/site/pinguinotutorial/home

Components

Material I built my board with:

1	Copper Board (VeroBoard or Stripboard or Perfboard)
1	PIC 18F4550
1	40 Pin IC Socket
1	USB Type B Socket
1	20MHz Crystal
1	220nF Capacitor
1	100nF Capacitor
2	22pF Capacitors
1	Push Button
1	Dip Switch
1	10k Ohm Resistor
1	7805 Voltage Regulator
1	9v Battery Connector
1	Female Pin Header
1	1N4001 Diode (or any other general purpose diode)
2	Rows of Pin Headers

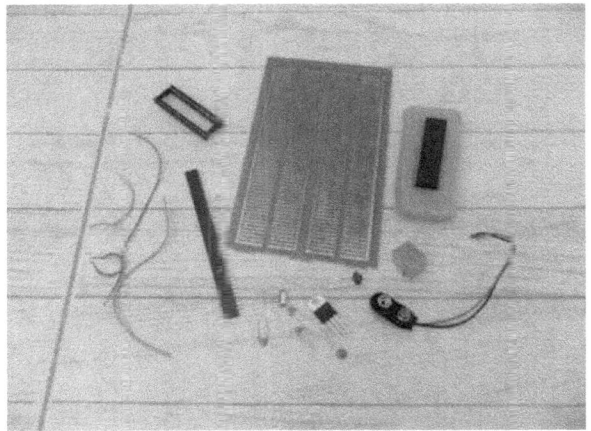

This circuit is the circuit diagram on the Pinguino website.

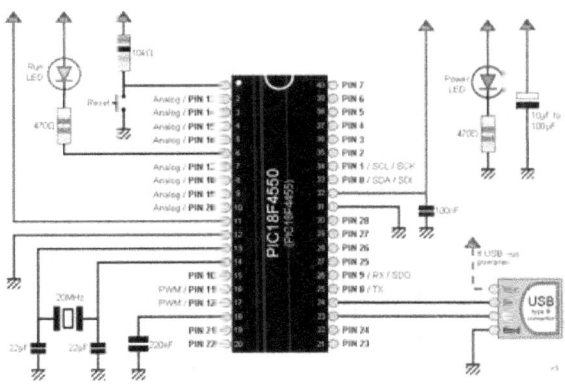

And this circuit is the one I've built in detail.

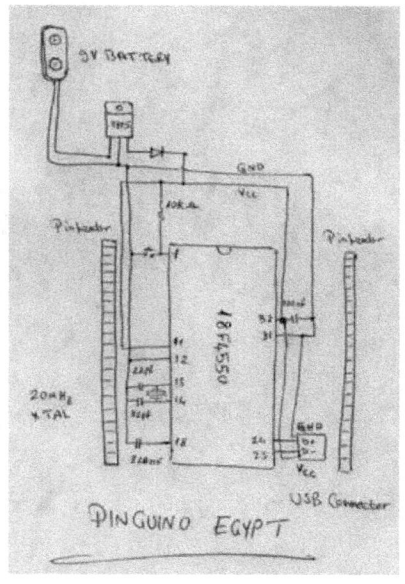

PINGUINO EGYPT

40-Pin PDIP

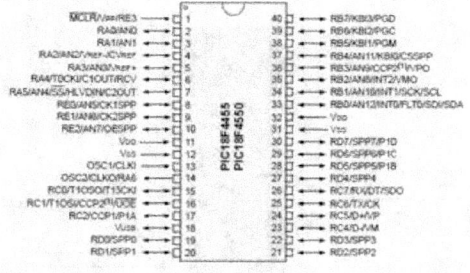

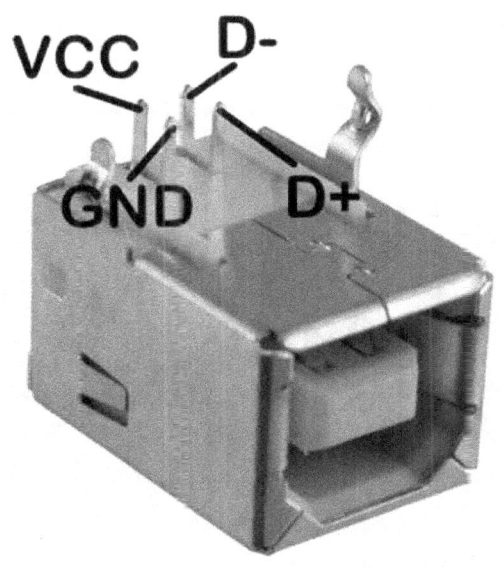

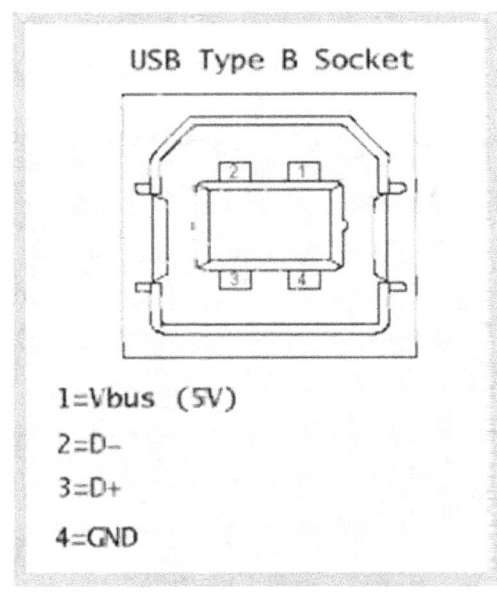

USB Type B Socket

1=Vbus (5V)

2=D-

3=D+

4=GND

Start Soldering

There are many instructables and other online tutorials for soldering.

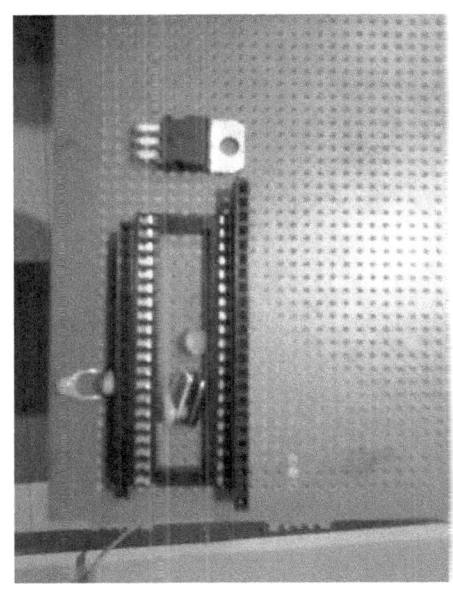

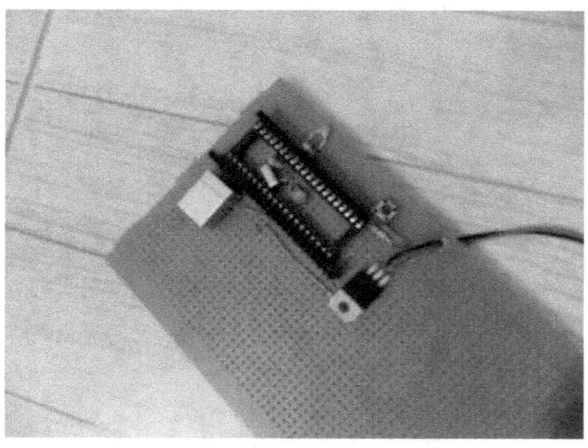

I used a large Vero board because I didn't know how exactly the circuit will occupy.

I started to solder the components on the board.

After I finished assembling and soldering, I used an architect's saw to cut the board into the size which the components actually fit nto.

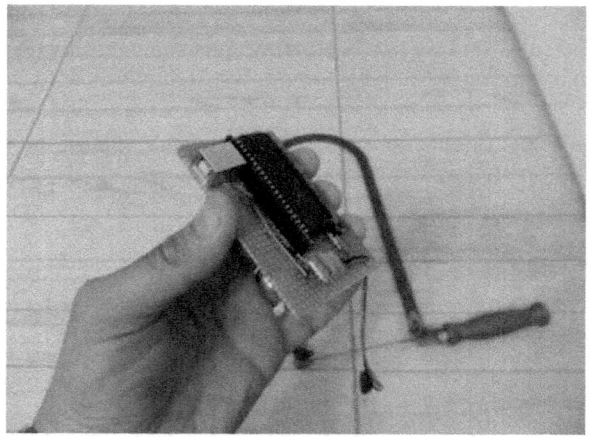

Install Pinguino Boot loader

The boot loader is the initial software that enables the Microcontroller to communicate with the PC through the USB port.

It also enables the Microcontroller to self update its software sent on the USB port.

This is the link were you can get the boot loaders of your circuit. In my circuit, I've chosen the PIC18F4550 Microcontroller with the 20MHz Crystal

http://pinguino32.googlecode.com/svn/bootloaders/8/usb-v4.x/hex/

The 18F4550 Microcontroller chip is a self programming Microcontroller.

This means that the Pinguino board can be used to update its Firmware to a new one making it capable of doing a new function.

Initially, the Microcontroller cannot directly communicate with PC using USB port. But it can be programmer as a normal Microcontroller.

You can program your chip on an external USB or Serial Microcontroller programmer.

You have to do this step only once. After this, when completing Pinguino, you will not have to use a programmer any more.

I installed the bootloader using a simple programmer I previously built. It is called JDM Programmer.

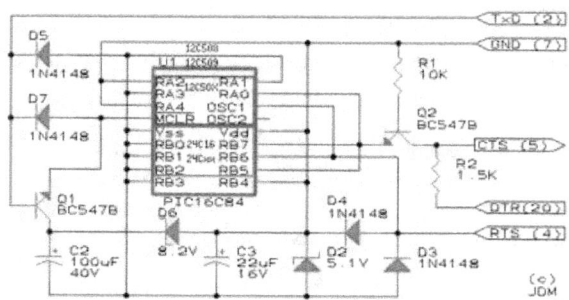

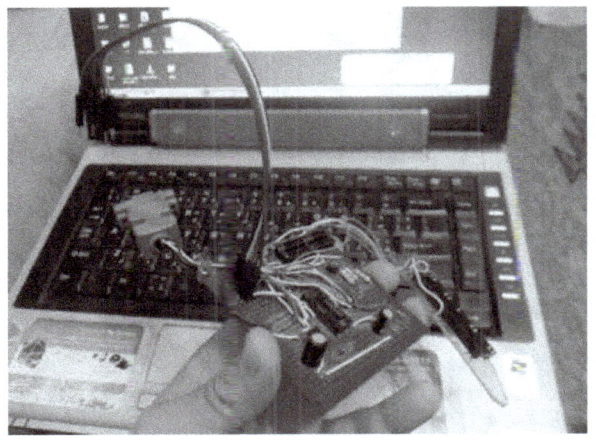

You can build it your self. It is simple and direct.

Here is a link for the Instructables of the programmer

http://www.instructables.com/id/18F4550-Programmer-in-Ten-years/

I used the PicPGM programmer with my serial JDM programmer to load the target bootloader file (Bootloader_v4.13_18f4550_X20MHz.hex)

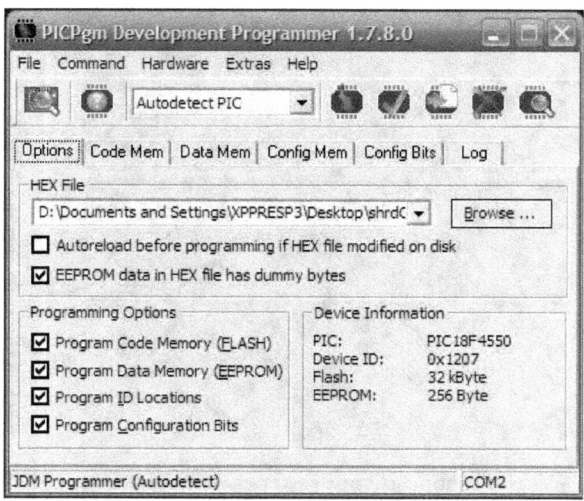

After successfully install the Bootloader Hex file on the Microcontroller chip, you can insert it into the Pinguino board.

Congratulations, you have completed the Pinguino board Hardware part!!

Pinguino IDE Installation

The Pinguino IDE is the software part of the Pinguino where you can develop code for your project, compile it and then upload it into your Pinguino board.

Here is how to install the Pinguino IDE on Windows Xp

http://wiki.pinguino.cc/index.php/Windows

It may look a little bit long. But believe me, it's not complicated.

It is simple and strait forward steps.

Here what I've done simply:

1- Download and extract Pinguino IDE for window form here

https://code.google.com/p/pinguino32/downloads/detail?name=Pinguino-
IDE-snapshot-i386-unknown-win32-20131209-
rev959.7z&can=2&q=

2- Download vcredist_x86.exe (for win32) --- If you encounter the
message of not existing msvc*90.dll in Windows,it has to be
installed from here

http://www.microsoft.com/en-
us/download/details.aspx?id=5582&WT.mc_id=MSCOM_EN_US_
DLC_DETAILS_131Z4ENUS22004

3- Download and Install Python from here:

http://www.python.org/ftp/python/2.6.6/python-2.6.6.msi

4- Download and install wxPython 2.8.12.1(unicode) for Python 2.6

http://downloads.sourceforge.net/wxpython/wxPython2.8-win32-
unicode-2.8.12.1-py26.exe

5- Download and install pyusb-1.0.0a2-py2.6

http://sourceforge.net/projects/pyusb/files/PyUSB%201.0/1.0.0-alpha-2/pyusb-1.0.0a2.zip/download

To install Pyusb module, you have to follow the next steps:

- Add python path to the PATH variable... For example   c:\python27

- Extract pyusb folder on any location of your hard disk

- Open the command prompt and change directory to the pyusb directory.

- Run the command python setup.py install

- Make sure that the Pyusb module is installed by python directory and look for site-packages folder... For example: C:\Python27\lib\site-packages if you can find USB folder inside then pyusb is successfully installed.

- You can now remove the Pyusb source folder.

6 - Download and install libusb-win32

http://downloads.sourceforge.net/libusb-win32/libusb-win32-filter-bin-0.1.12.1.exe?modtime=1174387137&big_mirror=0

(Don't discard this file yet. In some cases the library has to be installed twice to work.)

7- Download and install pyserial

http://sourceforge.net/projects/pyserial/files/pyserial/2.5/pyserial-2.5.win32.exe/download

Now, with a printer USB cable, connect your Pinguino board to the PC. You'll get the Found New Hardware popup message.

8- Download and Extract Driver from here:

http://www.hackinglab.org/pinguino/download/driver%20pinguino%20w
indows/driver%20pinguino%20windows.tar.gz

Now, from the folder of Pinguino IDE (..\PINGUINO\x4-easy-rev959\
) start pinguino.exe file.

That's it.

Now you can open any example or Arduino sketch, compile it and then upload it into your Pinguino board and start your application.

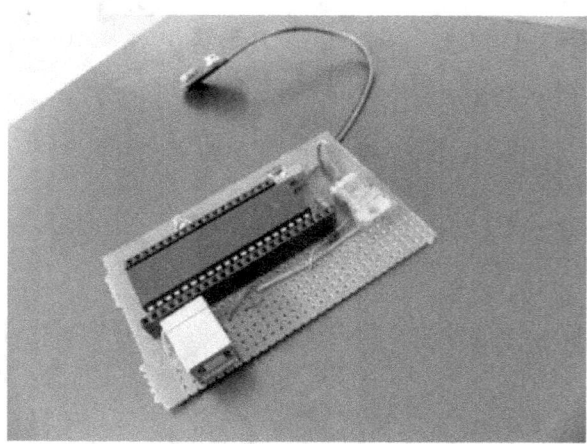

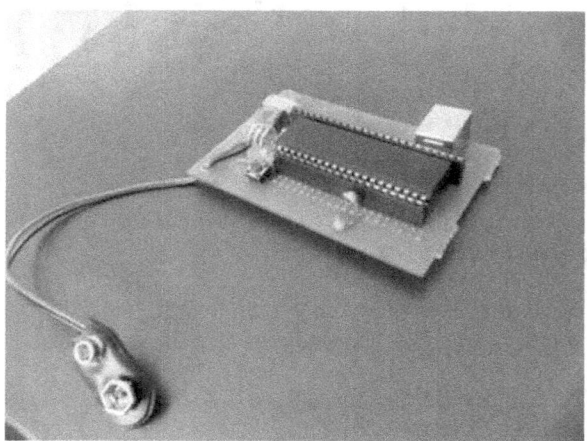

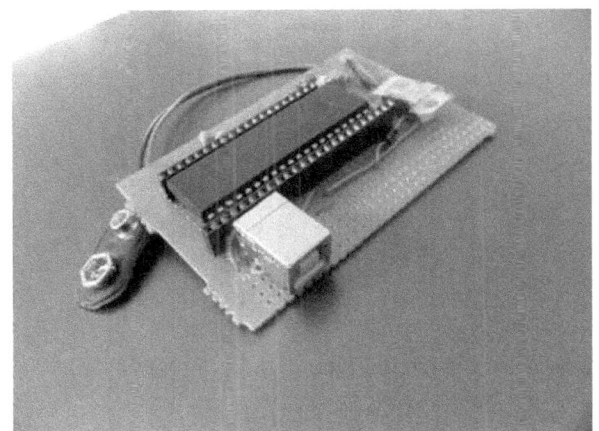

Hello World

The example here

http://wiki.pinguino.cc/index.php/Basics#Hello_World

Show you how to send text on the USB port using CDC emulated RS232

Just copy and paste the code

```
/*
---------------------------------------------
HELLO WORLD CDC
---------------------------------------------
*/

void setup()
{
  // put your setup code here, to run once:
}

void loop()
{
  // put your main code here, to run repeatedly:
  if (CONTROL_LINE) CDC.println("\n\r Hello World !!!");

}
```

And windows Xp will ask for driver software.

You can find it under Pinguino IDE Folder

..\PINGUINO\x4-easy-rev959\extra\drivers\CDC

Run **Hyper terminal** with the following settings:

Choose the com port which the Pinguino driver you installed has created. (Ex. Com 4)

Speed (Baud): 115200

Data bits: 8

Stop bits: 1

Parity: None

Flow control: XON/XOFF

You can read this article in Arabic

http://arabic-embedded-egypt.blogspot.com/2014/03/pinguino-egypt.html

Here is the article on my instructables.com page :

http://www.instructables.com/id/Pinguino-Egypt-PIC-Based-Arduino/

Thank you for visiting my blog. Have fun.

PSoC Rocks!!

Cypress PSoC 3 First Touch Starter Kit

I just received the PSoC 3 first touch starter kit. I new about it from the Cypress news letter I receive regularly. This is a free educational kit. It contains many sensors and shipped with sample codes for programs implementing all these sensors.

It contains thermal sensor, proximity sensor, capacitive touch panel, wireless communications module and my favorite sensor 3D accelerometer.

The first application I found as I opened the kit was PSoC Rocks. This is a manually scanned AirText display that is the same as the one I made using Microchip PIC 16F84. But this one is a more advanced application that uses the accelerometer in the direction at which the circuit is waved. It uses the POV (Persistence Of Vision).

When the acceleration threshold is reached, the scanned display is started. And it determines the direction of waving so that it doesn't draw the display in the opposite direction. Actually this is the same advancement I thought for my previous application.

I shoot this one myself as I started using the kit

For people who don't know about Cypress and PSoC:

PSoC (Programmable System on Chip) is advanced microcontroller architecture from PSoC which includes some digital and analog modules for simplifying designing Embedded Systems with digital and analog peripherals. Examples of digital modules (counters , timers , RS-232 , UART , USB ..) and analog modules (Programmable Gain Amplifiers , filters ...)

This is the actual kit I received

Microcontroller and Success: A true Story

At the year of 2008, I found an advertisement in Appliance magazine about a free kit from a Microcontroller company called Renesas.

I really wanted to get this development kit. It was dedicated to motor control. I registered for this kit but unfortunately it was available only in USA and Canada.

However, my email was added in the company's client's database. Next, the company started a design contest at the same year called Renesas HTS Design Contest 2008.

The company invited me to participate in this design contest. It was a world wide contest which included shipment of free development kit to each eligible participant.

First I was admired by the idea itself of receiving a free development kit for the contest. I wanted to have this kit to learn more about Embedded C programming and to learn the Renesas technology which was new to me until that time.

To be eligible to receive the kit, there was a short demo for the company's other educational boards over a Virtual Lab environment. Then you are asked few questions to qualify for receiving the kit.

I passed through the stages of qualification and waited to receive the kit. When I received it I started to develop my design right away.

The company made a very good idea for improving the challenge through the launch of a forum for the challenge.

Each participant posted the idea of his entry and provided some technical details about it . This idea was very helpful and motivational. Any visitor or another participant can see your posted details and get admired with it or comment on it.

My design was a Multichannel Oscilloscope. The idea was very simple. I made the analog signals read by the analog-to-digital converter of the Microcontroller M16C (on which the board is based)

The M16C has many A/D inputs but and there were 3 of them available in the challenge board.

The signals values were converted into digital in the Microcontroller and they were sent to PC based client software which incorporated the display of a virtual Multichannel Oscilloscope.

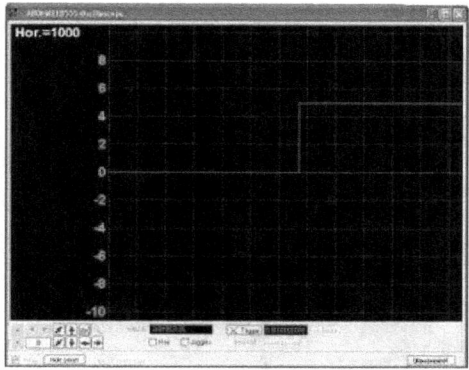

Virtual Oscilloscope

This is the Renesas HEW program window

From the PC based client, the user can choose the input channel to be displayed on the virtual oscilloscope and the rate at which it is displayed.

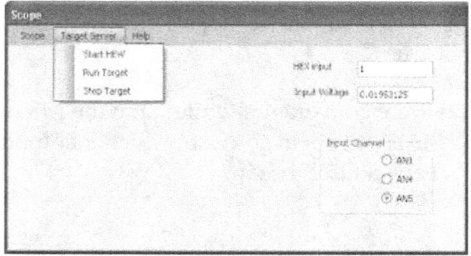

Program Control Panel

Before the last day of the deadline of the contest I submitted the source code for the Microcontroller side (written in Embedded C) and the source code of the PC client (written in Visual Basic.Net).

The results of the contest were announced on Renesas DevCon 2008.

I've got the 4th honorable mention prize for my entry "Multichannel Oscilloscope".

This contest was very challenging and rewarding. Thanks to Renesas who has helped me through this entire contest.

Location for the HTS 2008 projects:

http://renesasrulz.com/design_contest_archives/renesas_2008_hew_target _server_design_contest/m/mediagallery/276.aspx

Here is the location for my project:

http://www.renesasrulz.com/docs/DOC-1290

The link to the post in Arabic

http://arabic-embedded-egypt.blogspot.com.eg/2010/04/renesas.html

Gyro Horizon on Renesas RX62N Kit

In this post, we'll show an advanced kit from Renesas based on a modern microcontroller RX62N.

I received this kit by participation in Renesas Design Contest 2010. The kit has various types of sensors and an interface to the outer world that cannot be makes you wonder how to use them all in one application. For example, it contains a 3D accelerometer and a temperature sensor, USB, CAN, Ethernet and RS-232 interfaces and an alphanumeric LCD.

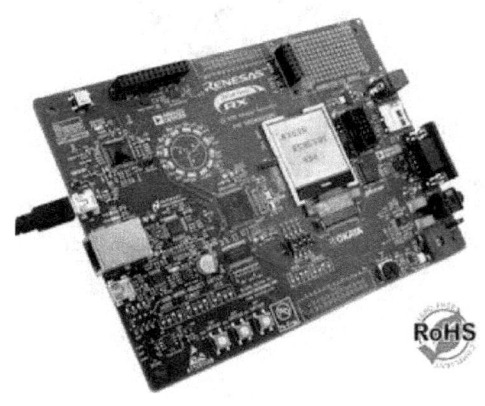

This is the Renesas RX62N Kit

The contest allows each contestant participate using only one application.

I wondered what application should I design, and I decided to design an ADI after being inspired by the Embraer 170 ADI (Attitude Direction Indication)

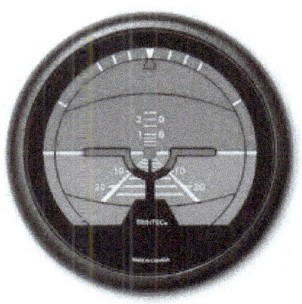

This is the ADI

The ADI is an important aviation instrument that helps the pilot controlling and the aircraft. It senses and indicates Pitch (Up and Down) and Roll (Right and Left) attitude of the aircraft.

The actual ADI instrument senses the attitude using sensors in the aircraft called Gyroscope. The modern types of ADI collects attitude data from an electronic device called Laser Gyro or Fiber Optic Gyro.

My version of ADI uses a built in sensor in the Renesas kit called Accelerometer

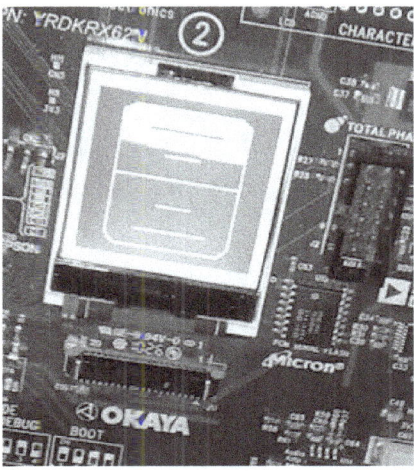

This is the running application on the kit (ADI appears on the LCD)

The Accelerometer is an electronic MEMS sensor that senses acceleration in three dimensions.

The Accelerometer senses the acceleration in three dimensions and then the RX62N microcontroller reads the data and then draws the output indication on the LCD on the kit.

This is the kit and the application on the LCD

Here is the contest entry:

http://renesasrulz.com/design_contest_archives/the_rx_mcu_design_cont est/contest_entries/m/mediagallery/260.aspx

Here is the project on instructables

http://www.instructables.com/id/Gyro-Horizon-DIY/

This is the link to the post in Arabic

http://arabic-embedded-egypt.blogspot.com.eg/2011/05/gyro-horizon-renesas-rx62n-kit.html

Arduino Temperature Sensor TMP01FPZ Interface

Now you have an Arduino board, what can you do with it?

Very simple temperature sensor project.

I got an analog temperature sensor from Texas Instruments TMP01FPZ .
Tow days ago I bought an Arduino UNO compatible board.

I wanted to make a simple circuit to use both.

Here it is.

Step 1: Components

- Arduino UNO and its cable

- TMP01FPZ Temperature Sensor

- Bread Board

- Some small wires

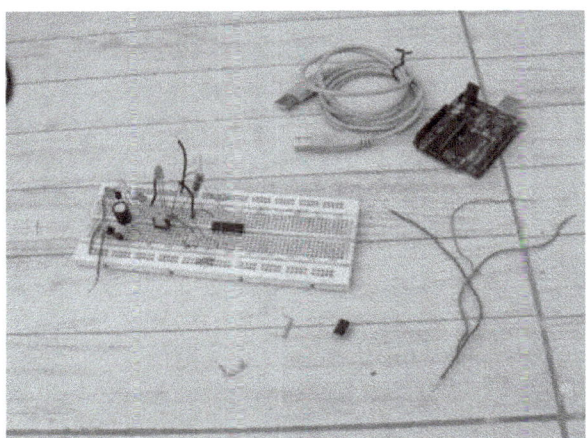

Step 2: Circuit

The circuit is very simple.

Just connect the Vcc and GND pins of the Temprature Sensor TMP01FPZ to the +5 v and GND pins of the Arduino UNO board to get it powered by 5 volts.

Then connect the analog out pin of the sensor to pin A0 of the Arduino UNO.

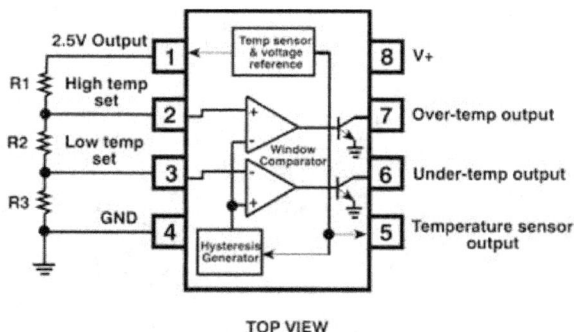

TOP VIEW

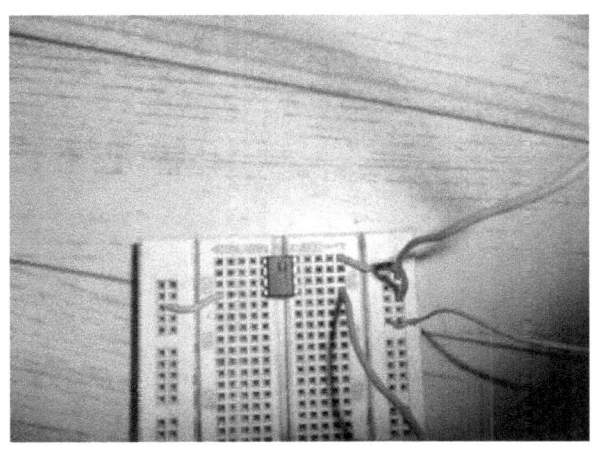

Step 3: Software

Configure the Analog input pin A0.

Read the analog input value to the variable **sensor** Value

Convert the analog input raw count into useful temperature degrees in Celsius according to the datasheet of the sensor.

Send the output degrees to the serial output on the USB port.

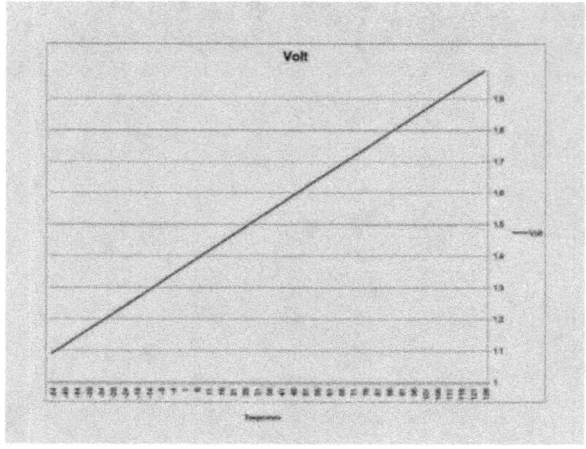

The chart represents the sensor response (output voltage) to temperature as described in the datasheet.

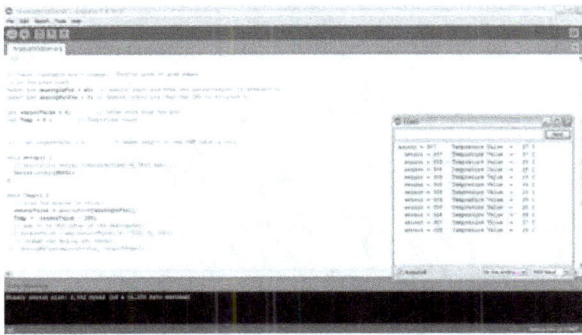

Read the output on the serial monitor.

Arduino code can be found here.

You can read this post on my page at instructables.com

http://www.instructables.com/id/Ten-Minutes-Arduino-Temprature/

You can read this post in Arabic

http://arabic-embedded-egypt.blogspot.com/2014/05/tmp01fpz.html

One Day with 1Sheeld

I attended an open day organized by Alex Hakerspace at Pluto and I was first introduced face-to-face with 1Sheeld.

With 1Sheeld Team

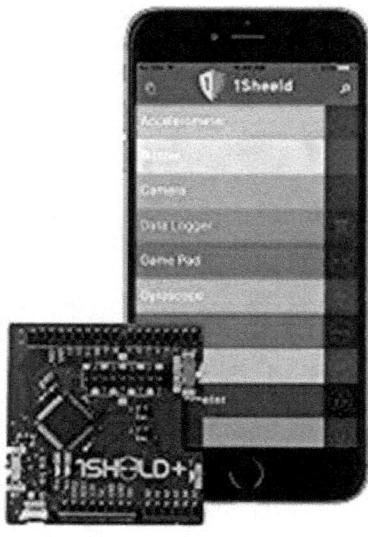

1Sheeld is an Egyptian product design and distributed by Integreight.

It simply replaces many Arduino shields with one shield and a smart phone running Android OS.

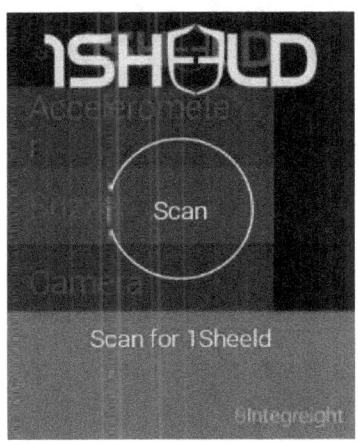

1Sheeld comes in 3 components:

- 1Sheeld circuit (1Sheeld Arduino Shield)

- 1Sheeld Android App.

- 1Sheeld Arduino Library

1Sheeld Circuit

The project is an open source Arduino Shield comes with an open source Arduino library and an Android App.

The team started a **Kickstarter** campaign and got the fund at 2013.

How does it work?

Get the 1Sheeld (or build it if you like)

Download 1Sheeld Android App.

Download 1Sheeld Arduino Library.

Install 1Sheeld on the Arduino board.

Run 1Sheeld Android App. on your smart phone and choose the shield you want.

1Sheeld Android App

Import 1Sheeld Arduino Library into Arduino IDE

You'll find that 1Sheeld Arduino Library was imported into Arduino IDE with all its Examples.

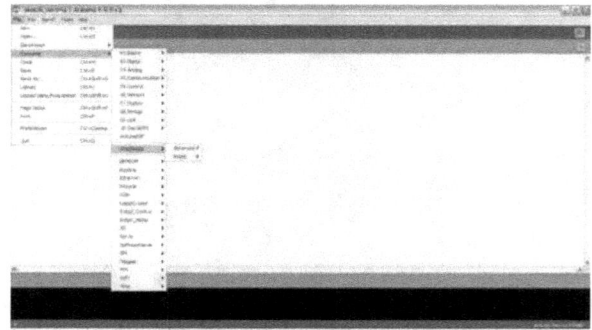

Links:

1Sheeld Website:

http://www.1sheeld.com/

1Sheeld Android App:

https://play.google.com/store/apps/details?id=com.integreight.onesheeld

1Sheeld Arduino Library:

http://www.1sheeld.com/1sheeld_arduino_library_v1.2.zip

1Sheeld Kickstarter campaign:

https://www.kickstarter.com/projects/integreight/1sheeld-replace-your-arduino-shields-with-your-sma

It is a great Egyptian Project. Thank you, 1Sheeld team.

Quadcopter building attempt

Today I started to build my first small Quadcopter. Some months ago I've decided to build my own small Quadcopter.

I've bought small Quadcopter motors from amazon.com for this purpose .

I've also bought a 3.7v   400 mAh Li-poly battery for the same purpose.

The four motors (2 pairs of motors, each pair have 2 motors which rotate in opposite direction) come with compatible propellers.

What I made today is a small step in building my Quadcopter. This step was attaching the motors into straight drinking straws as the first step in build my Quadcopter frame.

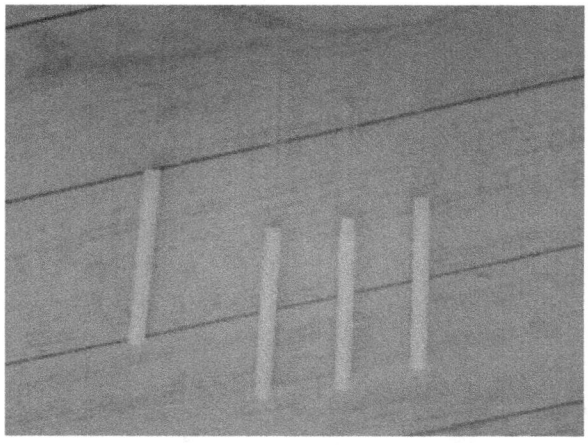

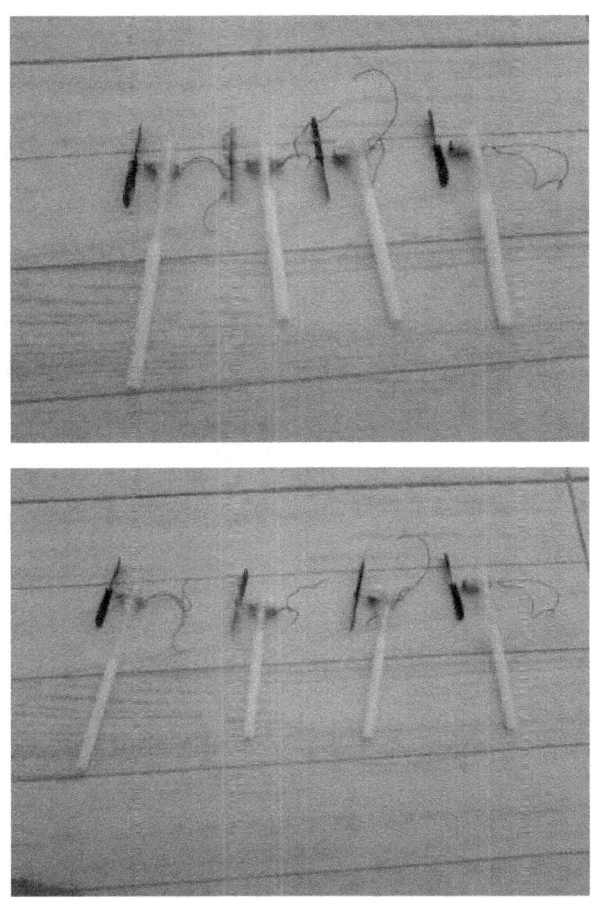

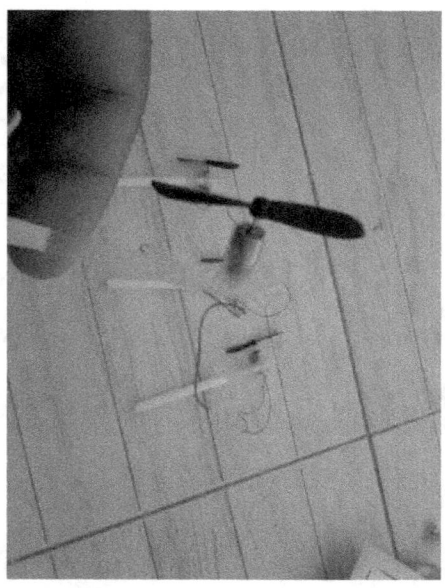

I also made something useful today. I've tried to attach a big Quadcopter propeller I've bought from amazon.com to a DC motor I got from an old CD-Rom player.

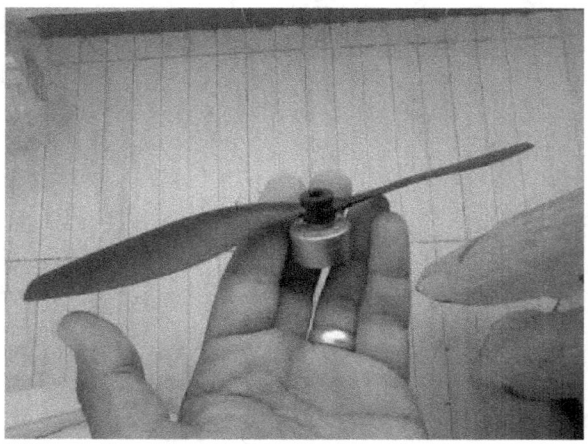

I supplied the motor with 12 volts from a DC supply.

I made this experiment to know how much lift is generated from the motor and the propeller.

Unfortunately, I got a smaller thrust from the motor than I've expected.

It is even less than that generated from the small motor with the small propeller. That's because the CD-Rom motor rotates in lower speed than that of the small Quadcopter motor.

To Read this post in Arabic

http://arabic-embedded-egypt.blogspot.com.eg/2015/09/blog-post.html

Quadcopter Flight Control Board

This is the first time to make a project that I post each step as I finish it.

The project I am working on these days is a self controlled Quadcopter.

Before, I used to build the project completely then I start posting it on my blog and on instructables.com

I am really enjoying it and feeling that posting small steps makes me more ambitious to make the next step and to finish the whole project.

In this step, I've built the Proteus simulation model for my Quadcopter board.

I really love Proteus and use it in all my embedded systems projects.

Here, I'll try a new approach. I want to drive the four brushless motors using the driver IC L293D using PWM signals generated from the PIC 16F777 Microcontroller.

I've another thing I don't know if it will work. I'll try to fly the Quadcopter using 2 axis accelerometer to measure the pitch and roll angels. I'll not use a gyroscope or an accelerometer in the 3rd (vertical) axis.

Here in the simulation software I intend to make an electrical simulation only (not aerodynamic simulation)

Here I used the ADXL206 2 axis accelerometer. I have no model for it in the Proteus software, so I used a couple of potentiometers to simulate the analog output from the accelerometer to the Microcontroller Analog input pins.

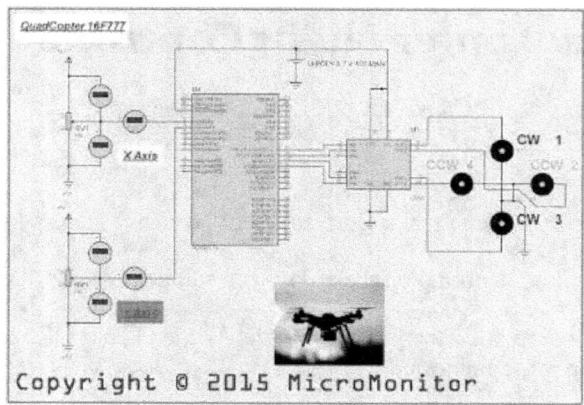

You can read this post in Arabic

http://arabic-embedded-egypt.blogspot.com.eg/2015/09/quadcopter-flight-control-board.html

L293D Four Motor Unidirectional Control

This is part f my project of building my first Quadcopter.

Today I wanted to try something I thought of. I wanted to control four DC motors using L293D IC.

The normal usage of the L293D IC is bidirectional motor control for two motors only.

That is because the IC is an H-Bridge circuit.

Since I am using the IC to control four motors for the Quadcopter, I am using the IC to control the motors in fixed direction.

No need to reverse one motor rotation direction (motor used to make lift)

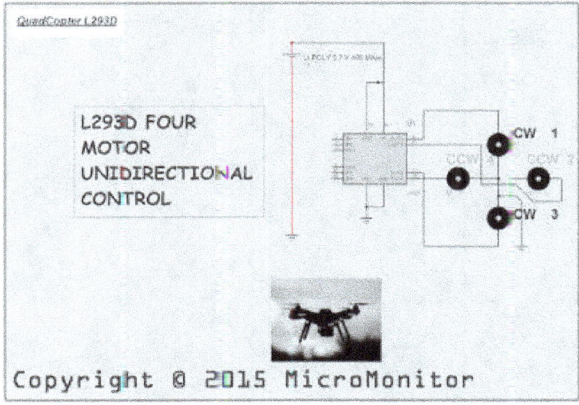

This is the circuit on the Proteus 7 simulator software

And this is the Link to the DSN file:

https://drive.google.com/file/d/0Byhyj_-
YLEr1UC05ZVM3V0Z4ekk/view?usp=sharing

You can read this article in Arabic

http://arabic-embedded-egypt.blogspot.com.eg/2015/10/l293d.html

PC Fan Wind Turbine

This post is not about Embedded Systems. But it is about a DIY project of mine. Old PC Fan to Wind Turbine.

I looked at the some old PC Fans I have and thought that they can be used as Small Wind Turbines.

The PC Fan is Brush less DC (ELDC) Motor. It can be converted to a generator in 5 Minutes.

The concept is simple. You can skip this part and start directly with the conversion.

The BLDC motor used here has a stator winding and a Permanent Magnet Rotor. The motor is supplied by 12V DC. But the magnetic field rotation is generated by electronics (Electronic Commutator). As the name implies, the electronics components convert DC into AC which makes the magnetic filed in the stator rotate.

The electronic commutation is achieved by a small IC.

To get the induced current from the motor used a generator, you must remove this IC.

This is what I'll show you how.

All what you'll need is this tool crocodile clip

First remove sticker on the back of the fan.

Then you'll find a small piece of plastic lock that holds Fan shaft secured, don't break it. Remove it with a crocodile clip.

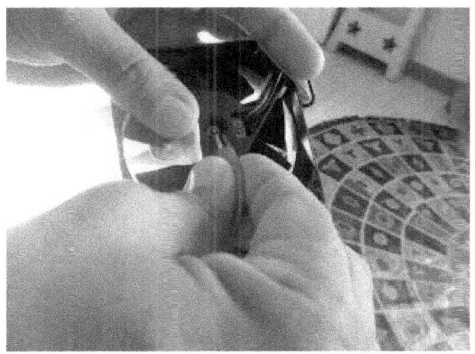

You can see 4 poles of winding connected in series and have only 2 terminals. To get the induced current, connect supply wires to those terminals and let the fan rotate.

With a solder iron, gently remove solder under IC pins and then remove the IC.

Remove this IC.

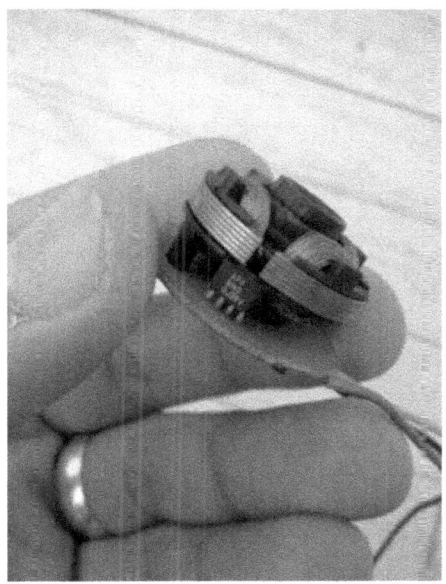

You can see that the winding is connected to the board at three points. One common and tow other terminals.

Remove the supply wires form there place to put them in the points where the tow winding wires are connected to the board.

Now, you have the stator ready to generate current. Now install the fan.

Assemble the fan in its place and lock it with the piece of plastic I told you about.

Put back the sticker.

Connect LED terminals to the Supply wires. Don't worry about +ve and -ve terminals, the LED will light if you connect it any way, trust me.

Roll the Fan

Here is the project on instructables.com

http://www.instructables.com/id/Old-PC-Fan-Wind-Turbine-in-10-Minutes/

You can read this article in Arabic

http://arabic-embedded-egypt.blogspot.com/2014/02/blog-post.html

http://idea-arabic.blogspot.com/2014/11/blog-post_26.html

TurbineOne - Basic Wind Turbine That Anyone Can Make

Have you ever wanted to have a wind turbine like this one?

Did you really like it so much that you always wanted to make it yourself?

This is my first working practical wind turbine. I really love green projects and renewable energy stuff. Last year I've made a small modification on an old PC fan to convert it into a small wind turbine. It had enough output power to light an LED. It was a huge project for me at that time because I've always wanted too much to get even little power form wind.

The huge number of people on instructables who successfully built different sizes and shapes of practically working wind turbines has motivated me to build my own next level wind turbine to have higher scale of power output.

That's where TurbineOne came from.

TurbineOne is my first practical power generating wind turbine.

I named it TurbineOne because I intend to build many other turbines.

I'll explain how I built it in the next steps.

I know when it comes to technical appeal, engineering calculations or technology practices TurbineOne is not very awesome.

Believe me; I'm not so handy when it comes to mechanics and using power tools.

Please make good comments and productive criticism.

Generator

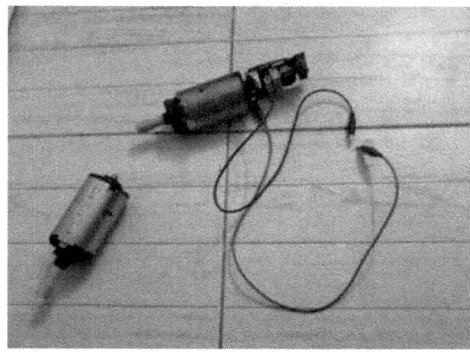

This is the most important piece of equipment for your wind turbine.

Actually, it was the first thing I started to look for when I decided to build my own wind turbine.

I thought to buy a DC motor from any hardware store who sells this kind of motor as a spare part for any appliance

(E.g. dishwasher, blender ... etc).

Then I found an old blender motor that has a permanent magnet inside it.

The motor generates electricity when it is turned by hand.

I measured the output and found to be nearly 14 Volts on the Voltmeter.

If you don't have an old motor to use as a generator, you still can buy a new one from Amazon, I got this one and it was useful.

Material:

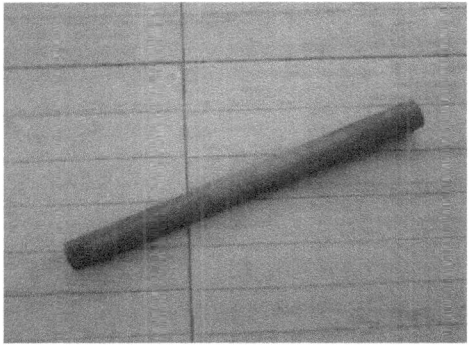

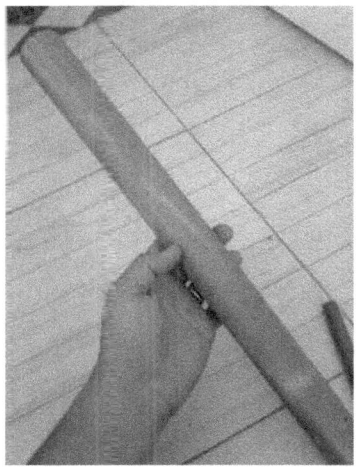

This wind turbine is 100% recycled. I got all parts from scrap and used stuff.

It took me a long time to collect some of the materials used for building it, but you can just buy them or be lucky to find them easier than I did.

Motor

PVC pipe --------- I found an old PVC pipe of suitable length to be used as turbine blades.

5 CD ROMs ------- I used old CD ROMs and DVD as wind turbine hub.

Fax paper plastic roll ----- used as a coupler between CD ROMs and motor shaft.

Old Plastic bottle ------ used as a cover for the generator

Some screws.

Some wires.

Old metallic rod used as a tower

Plastic tie raps

Tools

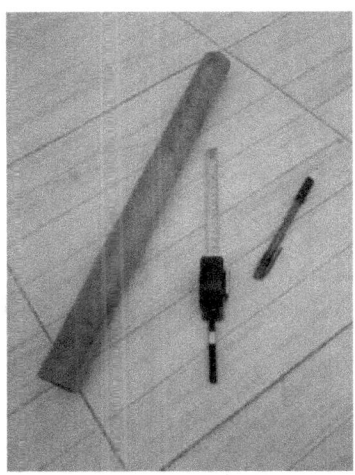

This project is made using fairly power tools. Please be careful when using this stuff.

- Saw

- Screw Driver

- Sand paper

- Pliers

Hub Assembly:

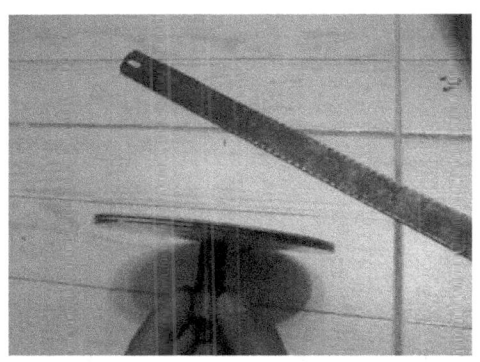

I started by the turbine hub.

Cut the plastic fax tube to 5 cm long.

I put the plastic tube around the motor rotating shaft.

Use the sand paper in CD Rom center to make the plastic tube fit into it.

Put CD ROMs and DVDs around the plastic pipe and motor shaft.

Step 5: Blades Assembly:

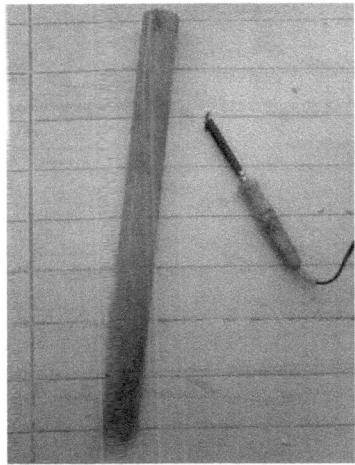

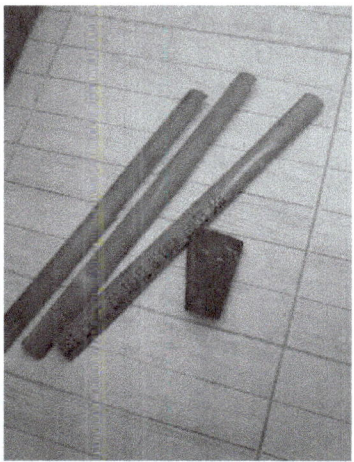

I wanted to cut the turbine blades into the usual turbine blades shape.

I really liked the idea of using PVC pipe as a fan blade. I got this idea from the internet.

But when I got the old PVC pipe I stated by drawing the fan blade on a template to draw it on the PVC pipe.

Then I couldn't get the tools to cut pipe in the fine shape of the fan blade.

So I've chosen to make the easy way and cut the PVC pipe into straight three equal pieces.

But how are these pieces going to generate rotational motion from the wind.

I decided to install each blade on the hub so it becomes nearly perpendicular to the hub and the round shape of the pipe does the rest.

Step 6: Turbine Assembly:

I installed the hub around the motor and secured them together using the fax paper plastic tube. Then I cut the extra piece of plastic from the small plastic tube. I drilled the metallic tube to install screws to fix the two tubes together. I installed the turbine with its pole on the roof of my apartment. It really rotates when the wind is fairly blowing.

TO DO: I'll make a battery charging circuit and connect a sealed Lead-Acid battery to make a steady supply power source.

I posted this project on the DIY website instructables.com

The post got featured on the website and in the weekly news letter.

I've participated at the Leftovers Contest on the website and got a runner prize.

http://www.instructables.com/id/TurbineOne-Basic-Wind-Turbine-That-Anyone-Can-Make/

You can read this article in Arabic Language from Here:

http://arabic-embedded-egypt.blogspot.com.eg/2016/02/turbineone.html

www.ingramcontent.com/pod-product-compliance
Lightning Source LLC
Chambersburg PA
CBHW051312220526
45468CB00004B/1307